Touch Screen Theory

Touch Screen Theory

Digital Devices and Feelings

Michele White

The MIT Press
Cambridge, Massachusetts
London, England

The MIT Press would like to thank the anonymous peer reviewers who provided comments on drafts of this book. The generous work of academic experts is essential for establishing the authority and quality of our publications. We acknowledge with gratitude the contributions of these otherwise uncredited readers.

This book was set in Stone Serif and Stone Sans by Westchester Publishing Services. Printed and bound in the United States of America.

Library of Congress Cataloging-in-Publication Data

Names: White, Michele, 1962– author.
Title: Touch screen theory : digital devices and feelings / Michele White.
Description: Cambridge, Massachusetts : The MIT Press, [2022] | Includes
    bibliographical references and index. | Summary: "Touchscreens are key
    elements of people's everyday lives but critical frameworks for addressing
    these devices and the associated promises of engagement and embodied
    experiences are still wanting. White proposes methods for studying
    touchscreens and digital engagements and expanding a variety of research
    areas, including studies of digital and Internet cultures, hardware, interfaces,
    media and screens, and popular culture"—Provided by publisher.
Identifiers: LCCN 2022000783 (print) | LCCN 2022000784 (ebook) | ISBN
    9780262544689 | ISBN 9780262372305 (epub) | ISBN 9780262372312 (pdf)
Subjects: LCSH: Human-computer interaction—Social aspects. | Touch screens.
Classification: LCC QA76.9.H85 W458 2022 (print) | LCC QA76.9.H85 (ebook) |
    DDC 004.01/9—dc23/eng/20220520
LC record available at https://lccn.loc.gov/2022000783
LC ebook record available at https://lccn.loc.gov/2022000784

10 9 8 7 6 5 4 3 2 1

Contents

Preface and Acknowledgments

In a seminar discussion about the touchscreen, I countered the popular image of the screen surface as clean and dry.[1] I narrated my experiences producing and using slick-with-sweat surfaces. I also noted changes in speed and movement, which occurred when my fingers came upon and were slowed down by embodied traces that were stuck on the screen. More recently, I found myself sweaty and sickened from excessive heat in the aftermath of Hurricane Ida and the total failure of the electrical grid. Challenged by a series of deadlines that a different organization, which shall remain unnamed, would or could not relax, I physically and mentally labored while crouched over my laptop and iPhone. This is a familiar and culturally shared telework position that requires further interrogation.

The associated devices were also lubricants and impediments. They were slippery because of the ways they "paired" and "unpaired" as I attempted to send files and find methods of accessing local hurricane news without using up what had become a precious resource—battery power. My body and the devices had different demands. I could use a renewable, solar-powered battery backup to run a fan and thus moderate my intense sense of embodied heaviness and dissipation. However, it was slow to recharge, even when labored over by my partner in the hot sun. The other option was to privilege the devices' requirements and thereby finish my work. Everything was wet—my body, the surface of the iPhone, and the touchpad and laptop case where it met my wrists and knees. My partner concernedly pointed out that perhaps the laptop was making me too warm. However, I couldn't sense the device because other feelings, including exhaustion and rage, predominated. I experienced an increased awareness of texture, including the abraded leather case of my phone, which was gouged with marks where my short nails had reached out and grabbed it. In such instances, my nails and the devices functioned as extensions and as reminders of the structured and sometimes misaligned conjunctions of physically touching and emotionally feeling.

Confronted with these and other experiences with stickiness, including the ways people are supposed to be connected to digital screens and desire them, I increasingly considered the limits of devices and my link to them. Companies and designers reference ties between individuals and devices as methods of eliding hierarchizing technologies in favor of visions of magical associations and pleasant feelings. Thus, Tim Cook's 2021 announcement about new devices, in a similar manner to Apple's narratives about previous products, insists upon an "experience" where "people love iPhone."[2] A related graphic morphs between and conflates the Apple icon and a heart and renders digital devices as part of people's intimate and impassioned feelings.[3] Such texts, as this book will indicate, also convey less agreeable digital relations and often dictate identity scripts that specify who is acknowledged and enabled to engage.

Some of these texts, in a different manner than sweltering against devices, allow me to think about digital objects. For instance, Conrad Bakker's "Untitled Project: Smartphone [Cracked Screen] [#2]" and the associated "painted sculptures of damaged smartphones with affected screens" depict instances where technological connectivity and personal intimacy have been broken (figure 0.1).[4] These screens are indeed affective, with their renderings of cracks and striated patterns. Such things, as my study of people's narratives about damaged phones indicates later in this book, are deeply felt. The felt "phone," even when portrayed by Bakker's art and situated in his hand, reminds viewers of the produced aspects of digital devices rather than their insistence on being transparent intermediaries that intermingle with bodies and become a form of skin. In one of Bakker's sculptures, the device's promise of "emergency" contact, which is blurred on the painted surface, is replaced by the emergency of the cracked and malfunctioning screen. Bakker's phones shift between notions of the digital window and representation and thereby question how these devices uncritically function as a "primary interface for engaging the world." Therefore, his work, like my sweaty and unpaired devices and some of the other texts that I consider later in this book, enable people to think with technological objects and to consider who is acknowledged and how the associated relationships are framed.

People as well as digital texts have been key to my studies. Some of the ideas about touchscreens outlined in this book, including my initial ideas

Figure 0.1
Conrad Bakker, "Untitled Project: Smartphone [Cracked Screen] [#2]."

about wet and slippery screens, were developed in conjunction with the attendees of the "Touch Screen Mediations: Intersectional Feminist Theories of Digital Devices, Bodies, and Applications" seminar, which I organized for the Society for Cinema and Media Studies (SCMS) Annual Conference in 2017–2018. A very preliminary version of my research on heart buttons and narratives, which is expanded and advanced in chapter 3, was presented at the SCMS Conference in 2016. I appreciate Jason Farman, Hollis Griffin, and Sarah Murray for including me in the associated "Mediating Mood: Experiencing the Interfaces of Social Media" panel.

More recent research that appears in chapter 3, especially my thinking on the link between hate and "smilies," was developed for the "When Individuals and Internet Systems Hate: Online Identities, Platform Affordances, and Technologies of Power and Harassment" seminar that I led in 2019 for the Cultural Studies Association Conference. I am especially grateful to Jennie Lightweis-Goff for her engagement in the associated imperative, and yet uncomfortable, subjects. It was also a pleasure to participate in wide-ranging discussions on affect with Susanna Paasonen and other friends at the Association of Internet Researchers conferences and related events. Students in my "New Media Theory: Touch/Photography/Screen/Theory" classes commented on literature related to this project and helped me to complicate the connection between theoretical texts about feelings and the functions of touchscreens.

As this brief list of events (and technologies) suggests, my ongoing critical endeavors have been supported by the generosity of the Carol Lavin Bernick Faculty Grant, Carol S. Levin Fund for Faculty Research and Creative Projects, Newcomb College Institute, Tulane's Committee on Research Fellowship, and Tulane University. My understanding of the Internet and related technologies and cultural practices continues to be illuminated by a summer seminar on literature and information technologies funded by the National Endowment for the Humanities (NEH), which Kate Hayles expertly directed at the University of California–Los Angeles. Her thinking on close reading and digital culture informs my practices and analysis in this text. Other seminar participants and friends, including Tara McPherson, contribute to my comprehension of digital culture and theory.

Unending thanks are also owed to the other friends and colleagues who sustained me through this research endeavor. For instance, the members of the Feminist and Queer Theory Reading Group, and especially the contributions of Jean Dangler, have been essential to my thinking. William Boddy, Mary Bryson, Courtney Berger, Alex Juhasz, and innumerable other friends have supported my academic career and proposed critical analysis of media and digital culture. Mike Syrimis and Michele Adams have, as always, offered steadfast encouragement. Mark Anderson continually located important texts and advanced my knowledge of technologies. He has also listened as I talked through my wide-ranging thoughts on this project. I owe a great deal to Noah J. Springer at the MIT Press for his thoughtful consideration of new media and critical guidance in developing this book. It continues to be a great pleasure to consider these issues with my feminist family. I have been able to write this work and other books because of Anderson, Pauline Farbman, and Stephanie White.

Introduction The Touchscreen That "fills the hand": Physically Touching and Emotionally Feeling Devices

My television and Internet provider coaxes individuals to "control" media options "from the palm of your hand."[1] The message encourages viewers to imagine the remote control in their hands, the weight and feel of the technological device against their skin, and the ability to regulate the associated technologies and screens. The referenced remote is designed with a texture that is velvety, textured, and pliant. To pick up the remote is thus to feel, including the movement of the individual's hand across the textured plastic and the visceral delight of the balanced weight and silky smoothness of the device. Small bumps on the back of the device and on top of some of the buttons indicate the places that individuals are supposed to position their fingers and touch. Button bumps and fingernail grooves articulate the shape and size of people's hands (and fingernails), the connection between hands and devices, how devices operate through hand movements, the tactile functions of remotes, and the ways individuals are supposed to feel about technologies and the associated screens. Through such arrangements, the manufacturer of the remote and my provider associate the physical television and other digital devices with notions of screen power and control. They also replace conceptions of remoteness and viewing with physicality and contact. Individuals who match my provider and the remote's embodied criteria are conceptually centered in front of the screen and articulated as ideal users. Such individuals, as I indicate throughout this book, may be directed to feel as much as or instead of viewing due to the emphasis on touch.

Apple's advertisements for mobile phones also highlight the links between screens, hands, and touch. Apple asserts, "With iPhone X, the device is the display. An all-new 5.8-inch Super Retina screen fills the hand and dazzles the eyes."[2] Apple's identification of its screen as the "Super Retina" evokes eyes and expert seeing, but the company also persistently references agile hands. Sensory experiences with iPhones are supposed to be fulfilling, with devices fitting individuals' hands, delighting their eyes, and matching people's desires. As Apple proclaims, its "vision has always been to create an iPhone that is entirely screen. One so immersive the device itself disappears into the experience." Thus, Apple identifies its iPhone 12 Pro as having "more screen than ever."[3] The company conjoins its narrative about innovation to visceral and body-oriented experiences, which it claims to facilitate. Apple thereby markets the ties between physically touching and emotionally feeling the screen in a manner that is related to and amplifies other manufacturers' promotions. I focus on Apple because of its articulations of technologies, especially its correlation of physically touching and emotionally experiencing screen-based devices, in the introduction and parts of chapter 1. I then provide a more interface- and viewer-focused analysis of the ways touching and feeling are correlated in the rest of the book.

Apple's advertisements suggest that computer and network elements are made into screens (and screen bodies). The company employs screens to conceal and displace the functions of a wider array of technologies. Cellphones and tablets function as moving curtains behind which people communicate, devices enable people to screen individuals and access information about them, and digital screens broadcast individuals' bodies and are connected to and collapsed with their skin. Screens, according to Apple's claims, fit bodies and lifestyles. I address such issues in this book by considering how women are informed that their bodies and fingernails are not a fit for iPhones (and how these women interrogate such embodied norms) in chapter 1; the correlation of the skins and abilities of men and devices (which attempt to make women into objects) in chapter 2; the ways social networks use heart buttons and icons to seem to physically and emotionally connect with individuals in chapter 3; how women autonomous sensory meridian response (ASMR) video producers render tactile addresses in chapter 4; and how touching and not touching generate emotional feelings during the coronavirus pandemic in the afterword. As I indicate in the title of this book, people's engagements with screens, and especially touchscreens, are linked to feeling. The graphic arrangement of *Touch Screen Theory* is designed to convey the ways screens shape sensations. Touchscreens also form and are formed by the individuals and texts that analyze devices. I thus offer readers a series of theories for considering touchscreens *and* for analyzing the related experiences of touching devices and being touched by screens.

This book and my analysis of touching and feeling are based on a multiyear study of blog posts, news articles, social media interfaces, technology forums, YouTube videos, and the related comments. This research is also informed by my archive of tens-of-thousands of examples and identification of the common themes and terms that appear in these texts, which I reference as part of my close readings throughout this book. I consider these sites as a means of highlighting how technology companies, devices, designers, and other individuals elide the constructed aspects of touchscreen devices and online settings by linking physically touching and emotionally feeling. People render technologies as accessible and unmediated methods of feeling by correlating women with devices and sentiments and by displacing women and other oppressed individuals' queries about how such technologies function. By focusing on the relation between physically touching and emotionally feeling, I recenter the bodies and identities that are empowered, produced, and displaced by digital technologies and settings. Part of this study thus emphasizes women and other subjects' methods of critiquing and reappropriating these devices. I assert that scholars' understandings and analyses of digital screens require an attention to the correlation of physically touching and emotionally feeling *and* the ways these purportedly unmediated practices efface representations, screens, and the associated technologies. Since digital screens, including mobile phones, have increasingly become media screens, communication platforms, and information delivery devices, my call to foreground conceptions of these screens, and articulation of theoretical and close reading methods, are designed to influence media research and humanities and social science scholarship more broadly.

My inquiries throughout this book suggest how digital screens are connected to an array of cultural practices and embodied sensations. The intermeshing of physically touching and emotionally feeling is evoked in Eve Kosofsky Sedgwick's *Touching Feeling: Affect, Pedagogy, Performativity*. Sedgwick argues that a "particular intimacy seems to subsist between textures and emotions," a nexus that is conveyed by the dual meanings of the terms "touching" and "feeling."[4] I address the ways feeling and seeing are associated with screen technologies and assist in producing viewers' positions.

I engage these visceral and mediated concepts, including the interconnection of tactility and emotions, through the literature on touching, feminism, media, and digital devices and online engagements. While I consider the contemporary literature on affect, my research emphasizes the critical analysis of touch and embodiment because these texts are more likely to focus on the mediated and representational aspects of embodied contact. My concerns about cultural conceptions of digital screens include the ways technologies are related to and made distinct from popular beliefs about and academic studies of gender, race, and sexuality.

People continue to associate touching, especially being touched, with disempowerment, closeness, intimacy, women, and femininity, and ordinarily relate culturally sanctioned looking to control, distance, men, and masculinity. The sociologist Mark McCormack describes how men limit their physical touch to activities that are identified as more aggressive, including contact sports and fighting, because of their fears of being described as girlish and gay, and thus deemed inappropriately masculine.[5] These correlations of ideal devices and positions with distance are renegotiated with touchscreen media. Men, as I suggest later in this book, tend to claim ownership of and an association with expensive and seemingly more expert touchscreen-based technologies. These connections also protect men from some of the features of touchscreens that are still identified as too feminine and intimate. Thus, journalists' concerns about the loss of touch during the coronavirus pandemic, which I consider in the afterword, are correlated with a purported lack of intimacy and the inaccessibility of women's bodies. This is meant to maintain men's position as the empowered touchers, or at least not bodies that are touched by other men. These articulated positions, which are built into the designs of devices and reporting, can be understood as a form of gender script.

Feminist Internet and science and technology studies, including literature on gender scripts, inform this book.[6] This includes studies of the sorts of worldviews and feelings that shape online engagements and technologies. I also continue my analysis from other research on the ways identities and sites are rendered. While the gender scripts literature tends to focus on how companies and designers articulate the users of devices, I advance these studies by addressing how individuals assist in the production of digital identities and norms. I connect people's production of gender scripts, as I note in more detail in subsequent chapters, to brand identities and the ways individuals are encouraged to contribute to companies' positions. My research thus demonstrates how the correlation of physically touching and emotionally feeling is designed to and does influence individuals and is a way of defining companies, devices, and online sites. I also show how the cultural association of touching and feeling asserts new sorts of unmediated emotional connections while producing normative and deeply structured forms of gender, race, and sexuality.

Apple suggests that its brand facilitates unmediated feelings when it prompts potential buyers to "Pick up the iPad" and "it's clear. You're actually touching your photos, reading a book, playing the piano. Nothing comes between you and what you love."[7] Picking up the iPad is supposed to displace the constructed aspects of the device in favor of the things that it represents and to put individuals in contact with and enable them to control the real (and a clear window and intimate, unmediated world).[8] There are cultural presumptions, as Lisa Gitelman and Geoffrey B. Pingree indicate, that "each new medium actually mediates less, that it successfully 'frees' information" from deficient media.[9] This effacement of the technology also displaces the ways bodies and identities are produced. As I assert throughout this book, it is imperative for scholars and the public to investigate the digital production and connection of identities, relations, and sensations because technologies are continually melded into the purportedly unmediated everyday.

The critical analysis and cases that I offer in this book are thus meant to provide readers with methods of addressing people's deeply constructed experiences with touchscreens and digital devices, especially their experiences of physically touching and emotionally feeling. For instance, I develop the term "to-be-touched-ness," which I derive from Laura Mulvey's argument about how women are produced as "to-be-looked-at-ness," as a means of identifying how digital settings constitute subjects, especially women's bodies, as touchable and women as compliant objects.[10] I further illustrate how the concept of tactile address, and its correlation with direct address and seeming to speak to a specific individual by using such terms as "you," allows for analyses of how women engage by referencing forms of communication that are culturally coded as feminine. I also note how women's constitution as to-be-touched-ness may seem to be an invitation and a tactile address, which further points to the gendered aspects of these textured engagements. Throughout this book, my interlinked employment of close reading and theorizing is intended to advance research practices. I firmly believe that critical methods for studying touchscreens and the associated cultural sentiments are vital to the analysis of broader societal practices and the continued development of such research areas as digital and Internet cultures, communication, gender and feminist studies, hardware and interface analysis, science and technology studies, and screen studies. While touchscreens are central features of people's lives, critical frameworks for addressing these devices and the associated promises of engagement and embodied experiences are still underdeveloped.

Touching, Feeling, and Affect

The names of digital devices and their features are designed to reference the experiences of touching and being touched. This is not a recent development. Lev Grossman directed *Time* readers in 2007, "Look at Microsoft's new Surface Computing division. Look at how Apple has propagated its touchscreen interface to the iPod line with the iPod Touch."[11] Grossman concludes, "Touching is the new seeing." Touch was also an aspect of earlier computing processes, including building computers, toggling switches, moving wires on plug-in control panels, and using keyboards. Grossman's comments apply to how individuals' hands and practices of physically touching are required. He does not acknowledge the persistent expectations that individuals see and locate visual material on screens. Of course, Grossman references and plays with the varied notions of looking and seeing as means of linking tactile digital experiences to visual engagements.

Scholars' studies of such narratives about digital and fleshy touch are part of recent research interests in the senses. Researchers' studies of touch emphasize bodily and intercorporeal experiences. Touch is thus understood as an experience that accentuates specific aspects of embodiment. Cultural conceptions of touch also link people and things in circuits of physical and emotional connection. Sociologist Mark Paterson notes that scholars often incorporate touch into investigations of the senses rather than independently addressing touch.[12] This may be because experiences of touching can be produced through other senses, such as tactile seeing and hearing. These multisensory phenomena include horror cinema representations and soundtracks, which viewers apprehend as skin tingling and embodied tearing. Individuals recognize depictions of extended hands as caresses. Paterson notes that touch is conveyed through pressure, temperature, and position. People may thereby associate being touched with the weight of a body or object resting against them, the sun heating their faces, being very close to other bodies, and a mobile phone vibrating and signaling that an intimate is messaging them.[13] While touch research can produce overarching notions of the body and sensations, I employ this literature and analyze touch because such approaches can assert the particularities of bodies, and thus challenge cultural perceptions and stereotypes.

People often associate touch with contact among physical objects and corporeal entities. Nevertheless, touch is ordinarily mediated, as I elaborate in more detail later in this introduction. Diana Adis Tahhan's study of intense connections and "touching at depth" indicates that touch is usually related to physicality and what can be seen but should be more broadly understood.[14] Tahhan employs the phrase "touching at depth" to identify a "felt relation and deep sense of connection."[15] She also proposes that touching at depth can be a method of displacing conventional conceptions of touch and offer a means of describing "intimate forms of touch and feeling." In my work, I employ this concept to consider the digital construction and connection of touching and feeling. In a similar manner to Tahhan, I supplant frameworks that focus on tangibility in favor of addressing the cultural entanglements of physically touching and emotionally feeling. As Tahhan asserts, notions of bodies as "separate, subjective and contained" suppress the ways people feel and connect. Related digital narratives about authentic individuality displace the ways individuals are told that they experience things and the commodification of their positions. It is worth considering the politics, economic value, and ethics of the feeling cyborg and of people who are persistently informed that technologies are part of their bodies and sensations.

Tahhan's notion of touching at depth, like the related literature that I detail throughout this book, is useful for theorizing individuals' engagements with and through devices. As Tahhan notes, acknowledging touching at depth and considering its theoretical and political implications can change how people conceptualize intimacy, embodiment, and feelings. The mediated, and thereby produced, aspects of touching at depth are underscored by Tahhan's narrative about the emotional connections that can occur when sharing communal spaces and viewing. She describes the enjoyable experience of collaboratively watching television with her family. This collective viewing has "sensuous" features "where sight and sound" link the family in the "depths of touch."[16] The family's shared presence and viewing are a "manifestation of intimacy here because they are a part of one another and share the same flesh."[17] Tahhan thus articulates the ways a variety of familiar and familial processes connect otherwise individual subjects.

The conception of touching at depth can unfortunately advance normative social structures. The idea that the space between people is inherently collapsible and there is an opportunity to be in touching relationships with everyone needs to be moderated so that it does not render or support notions of always-empowered subjects with permission to touch, individuals who are constituted as to-be-touched-ness and deemed to always consent, and global access. Touching at depth is still likely to operate within particular social groups and to be attached to cultural conventions and norms. Yet these connections also challenge notions that the body ends at the individual's skin, and thereby offer methods of considering more queer attachments. Touching at depth and the associated connections render shared flesh, which folds between and enwraps individuals and objects into intimate relationships. This folding and connection between bodies and skins occur when people wrap their hands around and extend their bodily forms into and conceptually through mobile devices, and when they are pressed against and enmeshed with family members (whether biological or queerly chosen), intimate partners, or hookups.

Tahhan associates home viewing with mutual forms of touching at a distance, or without contact between bodies, and at a depth, where people deeply feel things. Individuals also react to the more textured features of media, which in digital media include the cracked surfaces of touchscreens and the distorted images they deliver, sticky keyboards, and the pixilation of online videos. These forms of digital (and hardware) texture are associated with and can be considered in relation to Laura U. Marks's indication that such features as video graininess, soft focus, and camera positions that are close to the body evoke tactility and touch.[18] Marks also argues that representations of hands are not necessary for such haptic experiences, but viewing them evokes touch through association, including identification with individuals and/or with their hands.[19] So too do hand-pointers, which I discuss in more detail later in this introduction, amplify digital conceptions of touching.

Digital devices and representations connect the viewer's skin to the "skin" of media hardware and texts, including interface designs and thin plastic applications that are identified as skins, as I note in chapter 2. Thus, the tactile and resonant aspects of spectatorship, like the remote that I described earlier, are designed to intermesh media and embodied viewers. Vivian Sobchack's analysis of hand sensing and identification in the 1993 film *The Piano* (directed by Jane Campion) helps to explain such connections. She describes a scene where the view is blurred because the protagonist is holding her/our fingers in front of her/our face. Sobchack's fingers understood the film sequence, "*grasped* it," and tingled as if touching flesh.[20] She identifies the ways screens and texts render physical and emotional forms of skin contact. She also suggests how the experiences of understanding and holding things are conceptually intermeshed. Kevin E. McHugh similarly describes film as a "kind of skin-to-skin contact" in his study of "tactile-haptic cinema."[21] I employ these expanded notions of media-skin connections and the literature on skin when considering the gender coding of mobile phone surfaces in chapter 2. This includes the ways viewers' hands are intermeshed with and recognize touchscreens. As I also suggest in the afterword, digital forms of these connections have been welcomed during the pandemic and prompt concerns about viral spread. What these authors do not outline, and what I will consider in relation to the more proximal arrangements of viewers and touchscreens, is how physical closeness magnifies and fractures touching at depth and physically and emotionally feeling.

The corporeal and shared experiences that emerge from connections between things, such as Sobchack feeling through and becoming a film character's hands, are emphasized in humanities and social science texts on affect. Rather than being located in or generated by individuals, affect is often identified as the intensities that emerge from relationality. This includes associations among humans, nonhumans (that are corporeal), and inanimate forms. According to the psychology scholars John Cromby and Martin E. H. Willis, affects are intense engagements that arise during bodily encounters.[22] In Ruth Leys's review of affect theories, she argues that affect is often associated with presubjectivity and intensities that inform but are distinct from people's intentionality, cognitive processing, and decision-making.[23] In mapping out these distinctions, Ley also critiques the associated binaries.[24]

Scholars have employed theories of affect in considering the intensities that occur with digital technologies. The research on networked affect by

Susanna Paasonen, Ken Hillis, and Michael Petit underscores the role of nonhuman processes in affective encounters.[25] They argue that networked and online engagements are "underpinned by affective investments, sensory impulses, and forms of intensity that generate and circulate within networks comprising both human and nonhuman actors."[26] For instance, Paasonen uses notions of intensity and stickiness in her study of a Facebook dispute.[27] She demonstrates how affective forces stimulate online engagements and link people to platforms and the associated threads and groups. Sites can be conceptualized as "sticky" because prompts from interfaces, likes and other reactions, and cultural conventions encourage people to continually log in, respond, and be tied to interfaces because of their adhesive and tempting features. Sounds that indicate that a text or email has been delivered, news and app updates and notifications, and about-to-be live video feeds bind individuals to their devices and the associated sites and people.

Sara Ahmed includes stickiness in her expansive theory of "affective economies." She identifies how emotions stick individuals together or adhere them. Sticking produces the "effect of a collective" and "coherence."[28] This suggests how programmed features and online experiences bind people together and produce networks of feeling. Thus, affect generates the "*surfaces of collective bodies*," which include the constitution of the nation as a body, because affect is not located in a subject or object.[29] The touchscreen is also rendered and experienced as a body that is part of these circuits of feeling. Touchscreens are produced as bodies and rendered sticky through the emotional experiences produced by networks, the promises of connections with people and systems, and the persistent calls for responses. Sometimes screens are also sticky and repel associations, as I note in chapter 2, when bodily secretions and other forms of matter have accumulated on their surfaces. These traces record individuals' screen engagements. However, embodied liveliness is more generally associated with computer technologies by featuring screen-based images of smiling computers, describing computer hacks and malfunctions as "viruses," and putting computers to "sleep."

Hands and Touching Screens

Individuals are encouraged to mesh with mobile phones and other handheld and screen-based technologies. They are also prompted to identify these devices as fitting their bodies, and thereby suited to their interests.

Apple's iPhone X, as I suggest, is supposed to be shaped for and to fill the hand.[30] Apple thus works to amplify people's attachments to their phones and to assert that these devices are part of and an extension of the individual. In Heidi Rae Cooley's analysis of mobile screenic devices, she describes how people respond to tactile connections when their hands and devices mold to and interpenetrate each other.[31] This conjunction renders a version of Tahhan's touching at depth and Donna Haraway's networked and prosthetically enhanced cyborg, but not necessarily Haraway's politically resistant subject.[32] Cooley defines the links between hands and mobile screenic devices as a "fit," which may also be fitting. The term "fit" evokes manufacturers' and marketers' claims that phones fulfill individuals' interests; devices fit people's computing needs and hands; and individuals are agile in their use of technologies. When hands and devices mesh, hands also touch screens and individuals experience screens as touching.[33] As I suggest throughout this book, the practices of touching screens are rendered, identified, and felt as emotionally resonant.

Narratives about the intuitive fit between digital devices, sites, and finger digits are supported by people's employment of the terms "digit" and "digital." Media studies scholar Jack Bratich notes that "digital" refers to the "informational, virtual realm of ones and zeros but also to the fingers."[34] Shaun Moores, who studies media and cultural studies, identifies the "doubly digital" aspects of media, including the correlation of moving through online virtual environments and moving fingers and hands across touchscreens and other input devices.[35] The hand-pointer magnifies these connections because it is a digital representation, an element that activates computer processes, and a reflection of individuals' finger digits. It is one of the many features that correlate the underlying digital encoding and processing to particular people's fingers. As I elaborate upon later in this introduction and in other parts of this book, the hand-pointer conceptually recognizes and establishes whiteness and able-bodiedness as the norm. The hand-pointer is ordinarily a software and interface option in devices that are not touchscreen-enabled. Thus, the hand-pointer references touching when direct contact with the screen is not a programmed or preferred action. The hand-pointer represents the individual's hand "inside" and "outside" the screen and the hand of the individual connected to the hand of the device. It elaborates on familiar connections between people, including instances when individuals are holding or shaking hands. The handshake is also evoked in digital settings because it is a telecommunications term that describes the signals exchanged between devices when establishing a connection.

People feel further connected to other individuals, according to the mobile media research of Sarah Pink, Jolynna Sinanan, Larissa Hjorth, and Heather Horst, when placing their fingers on screens.[36] Individuals' presumptions about "sharing" hand positions may link people in hypothesized networks of devices and tactile versions of social media, which as I suggest in chapter 3 are elaborated upon with hearts and other mood emoticons and emoji. Apple references this notion of connecting, or even reaching through the phone, when informing consumers that its devices directly connect people to other individuals and the things that they love. Such texts promise to deliver the material and ideal world. Yet people's and manufacturers' articulations of touching as a series of physical actions and sensations should not displace mediated features. The seemingly conflicting practices of highlighting and ignoring mediation are in dialogue because individuals make hands into screens, see their bodies as screens, and identify the hand-pointer and the screen as versions of their embodiment, as I suggest throughout this book. Hands are also displaced from viewers' visual fields. This is in part a result of the ways device designs incorporate, or even absorb, hands and render them as stands, input devices, and pointers. More generally, hands are employed as notepads for handwritten reminders and tattooed representations and are inscribed with traces of people's labor in the form of calluses, scars, engrained dirt, and manicured fingernails. The associated fingernails screen the soft tissue under their surface even as they may display nail art and painted messages.

Hands and fingernails are often linked to specific gender and racial positions. For instance, Facebook represents its ideal participant as a white-collar (and presumably white) male worker by using an icon of a button-down-wearing and shirt-cuffed white hand. The designers of the Facebook icon and hand-pointer avoid cultural conceptions of feminine hands by not including fingernails in their depictions. Feminine conceptions of fingernails are also curbed by describing men's nail polish usage as "malepolish" and marketing such products as ManGlaze.[37] Touch is often characterized as feminine and more corporeal than such senses as seeing. Touch is evaluated in more positive ways when it is correlated with normative and able-bodied hands. The long philosophical tradition of identifying the hand, but not fingernails, as a key feature of humans is analyzed by Peter J. Capuano. He chronicles a "hand privileging" among philosophers and anatomists that the deconstructionist Jacques Derrida intervenes in with his concept of cultural "humanualism," or "*humainisme.*"[38] Derrida intercedes in the idea that having and employing hands is uniquely human and provides people with exceptional forms of handling and holding. Opposing thumbs, including their role in fine-grained manipulations, have also been classified as characteristics that make people markedly human. People sometimes link distinctions between humans and animals to gender, racial, and other identificatory hierarchies, as I note in more detail later in this book, and they use these categorizations to justify intolerance.

The philosopher Martin Heidegger, who claims an interest in "being-in-the-world" while downplaying the commonplace and communal, elevates the human when arguing that the "hand is infinitely different from all grasping organs—paws, claws, or fangs—different by an abyss of essence. Only a being who can speak, that is, think, can have hands and can be handy."[39] Heidegger conflates human hands with specific kinds of handicraft, tool use, speaking, and thinking. As Capuano suggests, there is a "deep etymological connection in German between manual grasping (greifen) and intellectual comprehension (begriefen)."[40] Such English phrases as "do you grasp this" also link holding to understanding. It has been suggested that Heidegger's articulation of the hand is correlated with his support of National Socialism.[41] National Socialists have employed the Roman or Fascist salute, with the arm extended and the palm facing down, to specify the characteristics and force of Nazism. Labor and activist movements have also employed raised arms and hands. Ahmed's feminist scholarship conceptually intervenes in normative and controlling arms when posing the willful girl and her raised arm as part of a history of feminist and labor resistance, including challenges to oppression. The force of corporations and online participants, as I suggest in chapter 1 and other parts of this book, too often refute feminist willfulness, protesting hands, and joined arms. I develop Ahmed's notion of the raised arm (and hand, and fingernails) as a means of underscoring women's critiques of the design of iPhones in chapter 1 and assertions of differently abled hands.

The ideal hand is instantiated by some digital interfaces and features. As I suggest in chapter 1, touchscreen designs continue to script able bodies and archetypal hands and fingernails. They also, as in the case of the

Facebook's thumbs-up icon and reaction button, specify hand and thumb gestures. I relate this to Tom Tyler's chronicle of how references to thumbs evoke intolerance. For instance, the "rule of thumb" and measurement of rods smaller than this metric started being used in the seventeenth century (or possibly earlier) to identify acceptable tools for abusing women.[42] In the Roman Coliseum, spectators' displays of thumbs were seen as support for killing the gladiators who lost matches. The relationship between these historical gestures, with spectators' thumbs in any position, and online interfaces' employment of thumbs-up and -down gestures remains unexplored. Tyler asserts a disability studies approach and argues for dismissing cultural classifications of opposable thumbs as distinct. He indicates that we should resist the model of an ideal hand and recognize varied hands, "gripping and grasping after their own fashion."[43] I engage with such critiques when highlighting cultural conceptions of and stereotypes about hands throughout this book, including notions of gendered and raced hands and fingernails. Hand typing, tapping, and gestures are central to digital engagements and could thus use more scholarly analysis. I provide additional frameworks for studying representations of and references to physical hands in digital settings.

The Hand-Pointer and Other Digitally Touching Hands

Hand-pointers are probably the most common computer representation of users. In this book, I expand my analysis of hand-pointers that appears in *The Body and the Screen: Theories of Internet Spectatorship* and other texts because hand-pointers frame our employment and understanding of a variety of digital technologies.[44] I also critique the hand-pointer because of the ways it is designed to assert that the white material body is present. Viewers often move their hands and change the associated position of input devices and pointers to engage in varied digital tasks. In some cases, interactions are achieved by manipulating the digital pointer until an image of the previously referenced white pointing and clicking hand, which indicates that something can be selected or manipulated, appears on the screen. These processes are designed to reflect the imagined position of the individual's hand. Since they are white, these hand-pointers link aspects of digital media to a white positionality and race aspects of computers, interfaces, and online sites. White hand-pointers continue to be the default, which can be changed in some cases by employing Apple's and Windows'

"Accessibility" system preferences. This associates other color options with disability. Whiteness is also privileged in the iPad advertisement that I have already referenced.[45] In one of the images, a white hand touches the screen *and* the textured and windswept hair of a young girl. The image underscores tactility, contact, and a white continuum because the child's own white hand holds a flower up to her cheek. This hand could be the hand of the device user who is "inside" and "outside" the screen. These raced hands are key features of how individuals are recognized and structured to touch, feel, and experience things. They are designed to elide the limits of white bodies and interfaces and to figure hands-on-relationships with elements that cannot be physically touched or carefully examined.

Representations of finger clicking and manipulating hands often appear in print and online instructional manuals for computers. For instance, Microsoft presents a series of white grasping and pointing hands in its development documentation for "Windows Desktop Apps."[46] In a related manner, Apple Developer's Human Interface Guidelines represents the "Closed" and "Open hand" for manipulating documents as white with a black outline.[47] Lines across the back of the hand reference the stitching on gloves, the strokes on gloves that cartoon characters wear, and the tendons beneath flesh. Apple, Microsoft, and other companies use these operating system and interface representations of hands to connect the hand-pointer to the material body. This configuration is also employed as a means of trying to resolve the problem of what Microsoft describes as the "weak affordance" of links. Microsoft uses a bold font to strongly advise developers that to "**avoid confusion, it is imperative not to use the hand pointer for other purposes.**"[48] The "hand pointer must mean 'this target is a link' and nothing else." Thus, the hand-pointer and touch are related to and activate the web address.

Contemporary hand-pointers are associated with the hand icons that Susan Kare developed in the 1980s as part of her work as the graphics and font designer for the Macintosh computer.[49] Her designs include icons for the open grabbing and pointing hands, desktop icons for drawing programs, and the pointing hand that is part of the Cairo font.[50] Kare's white hand-pointer is also related to animated cartoon characters' gloved hands. Early and continuing cartoon characters, including Minnie and Mickey Mouse, Bugs Bunny, and Pinocchio, wear white or light-colored gloves. These cartoon characters and their "white gloves, wide mouths and eyes, and tricksterish behaviors," as media studies scholar Nicholas Sammond observes, are part of a "long line of animated minstrels."[51] Blackface minstrels employed gloves and other classed goods as means of parodying and denigrating the aspirations of black people. As I have previously suggested, people also use hands to distinguish humans from animals and noncorporeal things. These connections between cartoon hands and operating system and interface hands are supported in online queries about "Who created the Mac Mickey pointer cursor?"[52] There are also indications that Apple called its mouse driver "mcky" in order to reference Mickey Mouse.[53] In making these references, technology companies and designers program minstrelsy and the associated conceptions of race, humanness, and animality into digital systems.

IBM started representing its personal computers with Charlie Chaplin's little tramp and his white-gloved hands in 1981. Chaplin is also referenced in IBM's 1983 PC*jr* brochure, which promises that the computer is an "easy one for everyone." IBM's claims about inclusion and simplicity are undermined by the brochure illustrations and photographs, which feature white families and individuals. Apple appropriated and represented Chaplin in the Macintosh "For the rest of us" ad campaign, which started in 1984. In Apple's representations, Chaplin's hand is positioned like the hand-pointer and is meant to evoke touching. This allowed Apple to reference its graphical user interface, suggest that its technology (rather than IBM's) serves everyone, and produce more affective connections. Apple continually correlates hands and digital devices, but it rarely addresses women and other individuals' fingernails. Instead, numerous news reporters and other individuals refused women's interrogations of Apple's design of iPhones and difficulties using the device with longer fingernails, as I note in chapter 1. Brand enthusiasts passionately refuted these women's critiques as methods of patrolling technological affordances and the people who are constituted as able users.

To-be-looked-at-ness and To-be-touched-ness

People's experiences being enabled to touch or prevented from touching, as my consideration of women's critiques of the usability of iPhones suggest, have gendered and spatial implications. Media theorists frequently address distant forms of visual engagement and looking, which are associated with men and masculinity. However, media texts also render intimacy and tactility. Mulvey is ordinarily associated with psychoanalytic considerations of the more distant and empowered male gaze, especially in her groundbreaking article "Visual Pleasure and Narrative Cinema."[54] The work of Mulvey and other critical scholarship on the gaze is concerned with how gendered, eroticized, and controlled bodies become visible within media and other texts. This includes the ways individuals look at, identify with, and are structured by screens and the associated visual representations. Thus, Mulvey argues that the cinema renders (white) women's position as to-be-looked-at-ness and "builds the ways she is to be looked at into the spectacle itself."[55] Aspects of media representations, including the lighting of women's faces and bodies, the textures of women's clothing and hair, and the soft focus of scenes that depict women, heighten the associated gendered structures and facilitate heterosexual men's erotic interests.[56] This suggests that the representational structures that render women and femininity as to-be-looked-at-ness also emphasize the haptic and to-be-touched-ness. Elizabeth S. Leet briefly makes this connection in her analysis of objectification in medieval verse. Leet chronicles how "textural details and allusions to adornment invite" men's interests and desires.[57] The organization of femininity as visual and to-be-looked-at-ness is combined with what Leet describes as the haptic aspects of texts and "to-be-touched-ness." The cultural constitution of women and other disenfranchised individuals as to-be-touched-ness also indicates that men can and should touch these subjects, and it makes women and other oppressed people into objects.

In this book, I indicate how contemporary versions of to-be-touched-ness have been developed and individuals have been further trained in gendered presumptions through such interface features as hand-pointers, thumbs-up and pointing hands, and the pixilation of images that make representations of women seem soft and ready to be touched. The link between touching and feeling is also articulated through heart buttons and emoji, as I suggest in chapter 3, and the ways these representations are supposed to convey people's love of goods, sites, and other participants. While touch interfaces and physical enactments of touching provide people with some level of control over digital devices and representations, being touched is still correlated with feminine and queer feelings. It is at the site of these purportedly excessive sentiments that critical interventions into the replication of norms are needed. Otherwise, women and other disenfranchised subjects will persistently be directed to constitute themselves as to-be-looked-at-ness, including being surveilled, and to-be-touched-ness. They will also be perpetually informed that being the object of the touch is low and contaminating.

Digital devices and representations continue and sometimes amplify gendered articulations of women as to-be-looked-at-ness and to-be-touched-ness. For instance, Apple's "Portraits of Her" video, which advertises "portrait lighting," emphasizes textures and an interface that allows a man to touch and change pictures of a woman.[58] The gendered features of portraiture, including the correlation of lighting with the objectification of women, are underscored by the absence of a "Portraits of Him" video. The video depicts The Shacks' vocalist Shannon Wise singing "The Strange Effect." Her pixie haircut, sequined top, and the camera's focus on her slender and shapely body associate women with tactility and to-be-touched-ness. Wise and other women are rendered as a picture and framed representation, while her light-skinned male partner, Max Shrager, is the producer of this representation and the person who holds and controls our view.[59] At one point in the video, Shrager grasps the device, fingers Wise's representation with his white hand (which acts as a version of the white hand-pointer), adjusts her light effects, acts as our avatar, and asserts men's control of and ability to touch women. He thus tactilely addresses Wise and constitutes her as image. Apple justifies this organization by having Wise sing that she likes this effect and position. As I suggest in more detail later in this book, direct addresses have tactile components that are supported by images of hands. These engagements depict addressed subjects as touchable and to-be-touched-ness.

Shrager stands in for the company and the device user when he frames Wise with his iPhone camera and suggests that she is interested in her production as tactile image—showing her the pictures that he has taken of her. She also demonstrates some power in striding through varied environments and demonstrating a series of light effects that viewers are expected to read. These frameworks shape how individuals understand digital practices and technologies, as I suggest throughout the book. Digital texts structure people's experiences through contextualizing names and descriptions, design elements (including gendered color options and features), promotional promises, and references to intimacy, the body, and the senses. Such digital configurations and technologies can be understood through the humanities practice of close reading or textual analysis. Close reading is not a principal academic methodology for studying digital culture. However, the many sites where people engage with digital culture by analyzing its aspects in detail, which include Wise's foregrounding of strange effects, point to the centrality of commonplace forms of close reading, and thus the productivity of scholarly close reading in these settings.

Close Reading and Touchscreen Theory

Throughout this book, I outline the critical practices of close reading that can be employed in studying digital devices and online settings. I also underscore some of the reasons that such analyses are associated with and can enhance our understanding of digital technologies and practices. Close reading is a productive method of digital study because it is related to everyday technological engagements, including the ways people employ computers and online interfaces to analyze posts and memes, the bodies of influencers, and the veracity of Universal Resource Locators (URLs). Reading is also mandated in personal and programmed reminders to attend to unread emails and update social media accounts. Apple presumes that individuals closely read its texts and emphasizes ambiguous meanings, even though it promotes unmediated experiences. For instance, Apple's iPhone 12 prompts individuals to "Blast past fast."[60] People may get stuck on this conjunction of words, which asserts that the technology is speedy while evoking fun and past digital experiences. I employ Barbara Johnson's attention to such ambiguities as a method of textual analysis. Johnson also proposes a method of reading that attends to the "meanings and the suspensions and displacements of meaning in a text."[61] She thus is not focused on identifying an overarching narrative, a canon of eloquent works, or authorial intent.

The new media scholar N. Katherine Hayles references Johnson in her consideration of digital reading. Hayles also relates people's digital research processes to their emotional experiences and mobile and constrained hands. She equates digital access to "the feeling one has that the world is at one's fingertips."[62] When technologies do not work, she "feel[s] lost, disoriented, unable to work" and as if her "hands have been amputated." Hayles relates the physical position of hands placed on keyboards and other input devices to the emotional experiences that are designed into and derive from touching things. She also evokes a different connection between the extended hand and digital power (and breakdown) than Ahmed's theorization of the raised arm. As I chronicle in chapter 1, Ahmed proposes that willful girls and women can use their raised arms to separate themselves from familial and cultural control and maintain determination over their own bodies.[63] Ahmed also reiterates words and concepts as a method of doing theory and closely reading terms and events.

Hayles has chronicled the varied ways individuals read and analyze digital texts, including close reading and theoretical engagements. In Matthew Kirschenbaum's study of digital mechanisms, he asserts that such close reading practices provide exceptional explanations of devices and texts.[64] However, close reading, as I have begun to suggest, is too rarely used as part of Internet studies. Explications of how to employ close reading when studying digital devices and online settings are scarcer. Hayles provides an outline of the aspects of digital comprehension that consists of close, hyper, and machine reading. She defines close reading as the "detailed and precise attention to rhetoric, style, language choice," and other literary and visual features of texts.[65] In this book, I employ such detailed, and at times arduous, analysis of textual components to support my critical and feminist arguments. Hayles relates close reading to people's deep focus and acceptance of boredom as part of their study. In a related manner, the English scholar Nicole Shukin argues that reading is a form of time-intensive and exhausting labor, especially when trying to meet expectations about academic acuity and scholarly coverage.[66]

Individuals' reading experiences include what Hayles identifies as "switching focus rapidly among different tasks, preferring multiple information streams, seeking a high level of stimulation, and having a low tolerance for boredom."[67] This hyper-positionality is associated with diverse and fragmented self-representations, multiple windows and media, skimming, and engaging assorted digital devices. In online settings, close reading is supported by and juxtaposed with hyper and machine reading.[68] The close reader is often understood to favor one information stream or text as a means of analysis. Yet, as I suggest in this book, close reading can also be employed to reveal and correlate the similarities among different sites and processes. When starting to closely read sites, I recommend attending to site names, URLs, logos and other identificatory images, color choices and additional design elements, "about" and "frequently asked questions" pages, the ways readers are addressed and represented, options for personalization and posting, content and the ways it is arranged, emotional and other icons, and repeated terms and punctuation. Due to these close reading practices,

I have tried to retain the unique textual and graphical aspects of online content in quotes and consider such features throughout this book.

People's time-intensive and reiterative engagements with texts, including the elements that I have outlined, link close reading to theory. Jonathan Culler and others argue that theory is something that individuals actively do when they employ theory in their own work or think with theory.[69] Texts become theories when other individuals find them to be useful ways of analyzing such things as meaning, cultural production, societal beliefs, and embodied positions. Some people use close reading to do a form of theory and employ texts that were not designed for the purpose as a means of considering everyday digital practices. This includes the ways memes and other visual images are employed as techniques of interrogation and critique. In a related manner, Nanna Verhoeff's study of the dual screen organization of the Nintendo DS indicates how the design of the game console foregrounds the interface and thereby raises critical questions about how individuals employ and can theorize with screens.[70] The traces of bodily matter, including oil, spittle, and hair, that stick to touchscreens may also encourage people to do a form of theory when such residues foreground the screen surface and its structures and representations. Such visible and tactile traces can also prompt people's feelings about how bodies and devices are supposed to function, including cultural expectations about clean and contained forms.

Jordan Alexander Stein expresses an interest in what he can do with theory in his book on the topic, including how theorizing feels.[71] Sections on silly, stupid, and sexy theory foreground the intellectual, political, and embodied struggles people engage in and feel like they are failing at when grappling with theoretical texts. As Stein notes, the competitive stakes in performing and citing recognizable theoretical concepts may result in people needing to dismiss other individuals' competencies. In such cases, theory is too often correlated with white male bodies and the related, and deemed appropriate, sensations. Academic directives to employ serious and competent theoretical arguments may be part of normalizing attempts to displace the queerness of theory and control shifts between different registers and modes of thinking. Digital studies, including the forms of analysis that I do in this book, can engage in the queerness of theory by accepting the messiness and abundance of online narratives. Such methods should also interrogate, and often refuse, the scripts of manufacturers and devices by recognizing diverse constituencies. Queer research trajectories can also combine everyday online texts with theory and employ useful texts rather than ones that signal expertise. Digital researchers can further engage with Culler's conceptions of close reading by foregrounding the multifaceted features of texts and their influences rather than resolving internal conflicts and differences. Critics of my methods tend to contrarily suggest that there should be more textual examples to support my arguments *or* that there are too many and too varied forms of close reading. In such cases, scholars risk trying to straighten and clean up my research, which is poignant when I am critiquing cultural concerns about messes, dirty devices, and nonnormative bodies. Of course, my appraisal risks acting as a method of excusing unresolved writing and concepts that could be further explored.

Stein and some other scholars chronicle their visceral experiences in simultaneously enacting close reading and theory. For instance, Terry Eagleton's account of how to read a poem recenters the embodied and material features of literature, which include how texts articulate and render sensations. He expresses concern that student readers are no longer attentive to form, which he indicates is not only the practice of noting the ways texts are written, but also when they are "shrill or sardonic, mournful or nonchalant, mawkish or truculent, irascible or histrionic."[72] When individuals focus on these elements and experiences, feelings are addressed. My close reading attends to such crafted and felt features as the cultural beliefs, sentiments, identities, narratives, and structural components associated with devices and online sites. As a means of addressing these associations, I analyze how devices and online characteristics are connected to bodies and feelings.

Michelle Marzullo, Jasmine Rault, and T. L. Cowan consider scholarly ethics and their feelings as researchers. They are knowledgeable scholars but still "felt unprepared for the methodological-ethical challenges posed by the Internet as a queer research environment" and had "some gut feelings about the need for better understandings of disciplinary practices."[73] Their account, which is infused with versions of the term "feel," underscores the stakes in and sensations of researching, including frustration, confusion, and exhilaration. Feelings about the extended temporal commitments to studying texts, which I have already outlined, are also referenced in Rault's consideration of the connections between close reading and queer Internet research methods and ethics.[74] She asserts that while humanities scholars' ethical practices tend to remain unidentified, they include closely reading for long periods of time and developing a knowledge of and a responsibility for the associated texts. Such humanities scholars of digital devices and online texts thus offer fine-grained analysis that acknowledges digital authors, texts, and modes of thinking and feeling about the everyday employment of technologies. As Rault's inquiry suggests, close reading can ethically acknowledge people's diverse investments in devices, online sites, production practices, identity constructions, and collectivity building. It can also outline how feelings are structured and evaluated through interface designs and the responses of participants.

Marzullo, Rault, and Cowan's apprehension about Internet research practices are understandable. Academic guidance about Internet research ethics offers conflicting suggestions to recognize individual authors and groups through citation practices, to anonymize people's texts, and to avoid some forms of online inquiries. The guidelines of the Association of Internet Researchers pose a variety of ethical questions and research practices.[75] Its most recent guidelines also highlight the threats that feminist and other scholars of online cultures face, including online attempts to sabotage scholars' careers and threaten researchers. These efforts to silence some researchers and forms of scholarship, as I suggest in this book, employ and are designed to generate feelings and render feminine fragility. Women are addressed as gendered bodies and threatened with rape and other forms of harm that are correlated with cultural beliefs that women are always and inherently at risk. As the philosopher Ann J. Cahill notes, threats of sexual violence assist in the production of fragile feminine bodies and constitute women as violable.[76] It is thus the case that researchers are called to experience their bodies, feelings, and scholarly imperatives in fraught and normative modes that could use additional forms of critical inquiry and political resistance.

In this book and in my other research, I attend to how individuals are produced and threatened by online representations. I also consider the practices that feminist, queer, and anti-racist subjects employ to critique normative devices and online engagements.[77] My attention to representations and organizational structures is related to the kinds of research methods that I employ in this book and my indications, here and elsewhere, that my research is not a form of human subject research that analyzes people. Such distinctions are made by US institutional review boards and the Collaborative Institutional Training Initiative (CITI), which educates US

researchers on human subject issues. These organizations associate intervention and interaction, including online surveys and interview methods,
with human subject research. Despite such guidelines, there continue to
be proscriptions about the kinds of research methods and scholarly texts
that are appropriate for studying digital media. Scholarly assertions that
all Internet research is human subject research and that sources should be
elided, as I have previously contended, do not acknowledge the deeply
produced aspects of Internet settings, the affordances of systems, and the
humanities methods that offer expanded understandings of sites and practices.[78] Such elisions continue to rank scholars and research practices and
efface the critical questions and information that are promoted through
humanities scholarship, including feminist queries about how identity
scripts validate white heterosexual male users. The associated directives to
understand Internet texts as people and research practices as human subject
research should be cause for concern. The sites that I study, and some Internet research practices, render all representations as authentic feelings and
material bodies and thereby make it more difficult to analyze how these
digital representations function, how they instantiate normative identities
and beliefs, and the ways they structure and hierarchize feelings.

These gatekeeping practices also occur within digital studies and humanities scholarship. There continue to be incentives for researchers to ignore
nonnormative embodied positions, forms of production, and histories, and
to engage with previously canonized critical literature and research frameworks. For instance, research on femininity is often identified as not rigor-
ous enough and as not contributing to the associated field and society more
broadly.[79] Such evaluative distinctions tend to be enforced and scholarly trajectories determined through editorial and referee practices. As anti-racist,
feminist, and lesbian, gay, bisexual, transgender, queer, intersex, and asex-
ual (LGBTQIA+) analysis of academic processes suggest, white heterosexual
men and masculine-oriented scholarship are advantaged by review practices.[80] In this book, I employ a feminist framework and research an online
contestation that has not been included in the histories of or theories about
touchscreens, probably because it features feminist critiques, femininity,
and the association of touchscreens and fingernails. I believe that performing research and generating theories in these areas provide vital methods
of highlighting the value of feminist and feminine practices. Such queries
encourage people to recognize and rethink disenfranchisement.

Touch Screen Theory Chapter Outlines

Individuals write about their favorite touchscreens and manage critiques as methods of maintaining their identity, association with devices, and relationship to preferred brands. I begin chapter 1 with a study of early reports about iPhones, including a series of articles and related commentary about how they do not work well with fingernails.[81] I study people's negative responses to a woman journalist—Michelle Quinn—and to technologists who indicate that normative notions of identity, and thus gender and other scripts, are incorporated into the design of iPhones and capacitive screens. These women argue that iPhone scripts delimit the ways women can employ the devices. I outline the literature on gender scripts and define it as the ways technology producers and designers imbue devices with their worldviews, and thus articulate the functions of objects and the people who are expected to use them.[82]

People's comments in related news stories and technology forums also produce a gender script when they figure longer fingernails, and especially women's fingernails, as antithetical to the use of screen interfaces. Such texts depict fingernails as aesthetic and frivolous aspects of women's self-presentations, and thereby fail to address the ways women are expected to produce their bodies. While early and current popular writing about mobile phones ordinarily suggests that these devices are tools that equally empower everyone, I demonstrate how these texts about mobile phones produce bodies and identities that tend to disenfranchise women. The gender scripts literature identifies designers and devices as the producers of these narratives. However, I expand the literature and its utility for critical interventions by identifying the ways individuals who employ and comment on technologies and identify with specific brands usually espouse beliefs that are related to technology companies and designers. I engage with the arguments of related science and technology researchers about how users matter, which indicate that everyday people as well as inventors, manufacturers, designers, and marketers decide upon the functions of technologies.[83]

The articles and posts that I study render women and long nails as a problem. I thus acknowledge Ahmed's indication that when women point to a problem, like the design limitations of phones, that they are identified as the problem. I also consider the ways women reference their hands as a means of pushing back against these frameworks and related forms of oppression. Ahmed's writing on the willful girl, raised arm, and feminist killjoy allows me to study women raising their fingernails and extending themselves as methods of protesting screen design and enacting change.[84] I also employ the scholarship on gender scripts to indicate how digital devices and online sites extend the beliefs of and promise to prosthetically combine with and authorize the identities of designers, corporations, and normative individuals. However, as I argue, the term "extensions" also describes artificial nails and the ways women employ their longer nail length and conceptually raised arms to reach things and revision cultural categories. I continue to theorize how fingernails function as tools, and I employ Ahmed's theories throughout this book as methods of thinking about how the hand is structured by online interfaces and how the hand-pointer and other devices are designed to link the physically touched and touching hand and emotionally feeling.

Mobile, or handheld, devices are enmeshed with and thought of as skin, including human skin. Yet these conceptions of skin, as I argue in chapter 2, are often employed to validate hierarchical categories and norms and evaluate bodies, especially the ways individuals are classified according to gender, race, sexuality, and class. Companies and individuals employ the term "skins" to describe digital and physical wrappers and to associate these skins with specific identities. People also relate mobile skins and corporeal-technological connections to sensations. Individuals describe pleasurable feelings about and attachments to devices and uncomfortable engagements with imperfect screen technologies, including screens that are identified as damaged and otherwise contaminated. My analysis of people's concerns about dirty phones and the viral transmission of the coronavirus demonstrates how the mobile phone is also employed as a means of pushing away nonnormative individuals and already disenfranchised subjects. Media theorists note the ways the ideal viewer is produced as a means of articulating hierarchies and cultural norms, as I suggest earlier in this book and in previous research.[85] In a related manner, I identify how the production of the skins and bodies of the mobile phone and owner render normative users and specify how individuals should appear and act.

In chapter 2, I study the similar online formulations of touchscreens that appear in blogs, manufacturers' websites, news articles, and technology forums. I focus on individuals' distressed and seemingly indifferent posts about phone damage and filthy phones. This allows me to expand my analysis from chapter 1 about how notions of contamination, including dirty nails, are dismissively correlated with excess femininity, women's bodies, and blackness. My analysis is informed by academic considerations of the body, including body studies' affiliation with skin studies. Body studies addresses such things as the ways the body is correlated with gender and other identity positions, more likely to be recognized when it is normative, experienced through a variety of sensations and feelings, and deemed to be disgusting and too fluid, especially when it is associated with women and other dismissed subjects. I employ scholarly texts about skin, including theorizations by Ahmed, Didier Anzieu, Nicolette Bragg, and Naomi Segal, to consider how devices become conceptually entangled with individuals and are impressed onto their bodies.[86] I argue that the cultural association of mobile phones with skin intensifies the connections between these devices and corporeality and feelings. Companies, designers, and individuals also employ these linkages to legislate identities and norms.

Sites employ gendered conceptions of heart and love to render participants and circulate their sentiments. In chapter 3, I continue my considerations from previous chapters and analyze how online cultures correlate hearts and love with women and girls, femininity, excessiveness, and queerness. These associations influence people's employment of heart icons, reactions to them, and understanding of how individuals use hearts and are produced by them. This includes the ways heart icons magnify and change individuals' feelings. Yet hearts can also be productive elements of social media because their cultural connotations and portrayals of feelings encourage responses. Cultural conventions and many people's interests in pleasing others mean that such phrases as "I love you" and "heart you" are directives for addressed subjects to feel the same way and to respond, ideally with versions of the same phrase.

Etsy, Facebook, and Twitter, as I note in this chapter, use narratives about hearting and loving and the associated icons and buttons as methods of scripting individuals and their connections to sites. These frameworks thus encourage individuals to love sites and to enact versions of brand love and community. Due to the impassioned connotations of heart icons, people experience these symbols as intense expressions and respond with similar sentiments, including love and revulsion for sites, brands, and individuals. Some people, especially men, identify hearts as contaminating because they are associated with femininity. Heart icons can be part of the banal everydayness of online expressions, which is the antithesis of intense and specific online passions. Nevertheless, many individuals on Twitter expressed extreme negative feelings when the site shifted from using the star to the heart as a means of denoting favorites and likes.

I employ the continental philosopher Roland Barthes's theory of punctum, Steve Woolgar's research on "configuring the user," scholarship on gender scripts, and the literature on brand love and community to consider how heart buttons, emoticons, and reactions produce feelings.[87] I demonstrate how these scholarly frameworks can be employed to consider the ways sites use conceptions of heart and love to produce participants and broadcast their feelings. Woolgar has shown how companies, technologies, and designers organize individuals and configure the ways they engage with products. Barthes's theory of intense bodily experiences with photographs also suggests how individuals are configured. His conception of photographic punctum, which has been identified as a form of affect and feeling, provides me with a way of considering how people respond to heart icons. People experience punctum when viscerally engaging with the unexpected aspects of photographs. Barthes's theory articulates the deeply embodied, unshareable, and queer experiences that individuals have with representations, such as the ways hearts directly address individuals with extreme sentiments and spot the surface of sites and messages with excessive feelings. Hearts are also normalized features of texts, which are thereby sometimes distinct from the ephemeral and uncontrollable aspects of punctum that pierce viewers and render intense feelings.

Intense embodied experiences and what is often popularly referred to as pleasant "brain tingles" are also produced by autonomous sensory meridian response vloggers.[88] In chapter 4, I study how these online video bloggers self-represent as ASMRtists (ASMR artists) as a means of underscoring their own skills at rendering feelings and tactilely addressing viewers. For instance, they produce tactile connections by reaching out as if they are touching viewers' faces and brushing against their hair. Visual devices, such as angled planes and extended hands, further link ASMRtists to viewers. Related forms of direct address, as scholars suggest, are employed in television and other media forms.[89] Individuals, companies, and sites' online forms of direct address constitute intimate bonds between people. They seem to speak directly to individuals by communicating with "you," creating visual settings that continue viewers' spaces, mirroring viewers' positions, and reaching out as if to touch and connect with viewers. Since ASMRtists rely on these addresses and the production of feelings in viewers, I develop the concept of "tactile address," which I mention in other parts of the book, as a means of considering how ASMRtists advance the functions of direct address and the video form. I also point to how my analysis of direct and tactile addresses provides additional methods for thinking about online sites, digital devices, and embodied feelings.

I focus on ASMRtists' YouTube screen-tapping videos. In these videos, ASMRtists touch, tap, and scratch digital screens and other objects in differ- ent rhythmic sequences. They position their screens to mesh with viewers' screens and compare their cellphones with viewers' hardware. They thus appear to touch viewers along with their screens as ways of rendering tactile addresses. ASMRtists' frequent dialogues with viewers in comment sections, including tactile addresses to individuals and the use of heart emoticons and other endearing expressions, are designed to amplify the feelings con- veyed by ASMR videos and experienced by viewers. These sensations are referenced in ASMRtists' and viewers' considerations of the pleasant visual, tactile, and sound-producing aspects of fingernails. They thus suggest that nails are ideal instruments and physiognomic features for screen engage- ments. This is distinct from some early articles about iPhones, which I con- sider in chapter 1, that articulate the challenges women experienced with the shift from resistive to capacitive devices because of their fingernails. While women with fingernails are ordinarily dismissed as being too aes- thetically self-involved, I argue that some ASMRtists offer a positive stance on this form of women's embodiment by emphasizing their long and active nails in tapping videos. In doing this, ASMRtists engage and disrupt cultural associations of women as to-be-looked-at-ness and to-be-touched-ness.

Throughout this book, I have considered the ways the body is engaged in and rendered as physically touching and emotionally feeling. People's corporeal positions are understood as screens and represented as being in contact with other bodies through screens. Bodies are rendered as screens when they hide devices, act as frames for mobile phones, convey facial expressions, and project information through fingernails and other embod- ied features. Screens are also re-visioned as bodies when they are described as living and dying and identified as friends and other intimates. These articulations of embodied screens and touch have been met with dismay because of the severe acute respiratory syndrome coronavirus 2 pandemic, which I will refer to as the "coronavirus pandemic" throughout this book, and the associated coronavirus disease, which I will describe as "COVID-19," as it is identified by the World Health Organization and other institutions.[90] However, forecasts about the death of screens and physical forms of engagement, as I suggest in the afterword, are indebted to larger cultural frameworks. This includes the continued cultural investments in digital forms of physically touching and emotionally feeling that I outline throughout this book.

In the afterword, I focus on ASMRtists' and journalists' accounts of touch during the coronavirus pandemic. My analysis of their narratives demonstrates how physically touching and emotionally feeling are correlated with socially distancing, emotionally feeling, and being characterized as unfeeling. Cultural evocations of not feeling include concerns that people are no longer emotionally engaging with other people because of digital devices, indications that people are socially distancing and avoiding interpersonal touch during the pandemic, and dismissals of women's agency and feelings about unwelcome touching. An unfortunate part of this formulation is the ways reporters identify #MeToo activists as at fault for and as a kind of prehistory to people's resistance to touch during the coronavirus pandemic. For instance, journalists propose that #MeToo has resulted in people being afraid to touch other individuals. In doing this, they fail to consider how consent and embodied agency are emphasized through #MeToo. I indicate that the journalistic replacement of #MeToo with an overwhelming longing for casual touch resituates women as to-be-touched-ness and as compliant objects. Touchscreen theory offers methods of addressing the cultural and personal implications of such structurations of escalated moods, including sentiments about not feeling. It also provides me with ways of thinking about how people enact and write about physically touching and emotionally feeling while these experiences are missed and disputed.

Throughout his book, I demonstrate how the critical literature on touching and being touched, including the writing of Tahhan and Barthes, offers ways of theorizing digital devices and the production and correlation of physically touching and emotionally feeling.[91] I also employ critical literature as a means of developing theories of tactile addresses and to-be-touched-ness. Feminist scholars have indicated that Mulvey's consideration of how women are constituted as to-be-looked-at-ness by classic Hollywood cinema is related to more general cultural expectations about women, including designations that young, normatively sized, and light-skinned women are desirable and should make themselves available.[92] The cultural structuration of these women as to-be-looked-at-ness contributes to their construction as to-be-touched-ness and the dismissal of other feminine subjects. Touchscreens and other digital media's emphasis on hands picks up on and escalates cultural conceptions of normative femininity and women's to-be-touched-ness. These technologies and related social practices also portray women as disruptive when they refuse or are unable to become acceptable bodies, and when they demand bodily autonomy. Thus, Ahmed's evocation of the raised hand and women's employment of #MeToo to identify experiences with nonconsensual contact are methods of critique and action. When women even momentarily join hands and point to the inequities of device conventions, as I suggest in chapter 1, brand consumers identify such raised hands as unreasonable disruptions. This suggests how the action of raising hands in protest can also stall normative everyday processes. The raised hand and #MeToo, as I indicate in the afterword, are theories that proffer other ways of reading culture and enacting relationality.

1 The "iPhone fingernail problem": The Gender Scripts of Capacitive Phones

The MacRumors site indicates that it "attracts a broad audience" and an "active community" of individuals who are interested in Apple products.[1] The associated MacRumors icon is a sliced apple with a seeded core, which evokes a face and thereby connects the Apple brand to its enthusiasts. This envisioned connection between products and embodied experiences is emphasized and disrupted when katie ta achoo inquires, "Are Long Fingernails Compatible with iPhone?"[2] She asks if she can actively use the iPhone, and if members will engage with her query. She also asks about women's experiences, which "girls may face that the average male wouldn't." Then katie ta achoo advises readers that she has been successfully texting with a mobile phone. She relates feminine experiences to technological proficiency and incorporates women into the MacRumors and Apple brand communities. However, HotdogGiambi changes her query to **"Are long fingernails compatible with LIFE?"**[3] He thereby threatens to make her undead or inhuman and rebuffs her attempts to be included. Hotdog-Giambi also challenges katie ta achoo about her technological engagements and skills. He performs a form of mansplaining when communicating for her. HotdogGiambi suggests that fingernails are a "ridiculous cultural stereotype perpetrated on women." He directs katie ta achoo to "Break free of the Matrix and trim those talons and stop painting your face like Bozo the Clown." With such texts, HotdogGiambi uses a liberation narrative to dictate what is acceptable for women. He blames women for their cultural position and for not resolving their oppression and self-representations.

Women are also dismissed and continue to be subjugated because they are presumed to be frivolous for engaging in beauty culture and monstrous because of their physiognomy. HotdogGiambi suggests that breaking out of the Matrix will free women. However, online men's groups commonly reference the film and the red pill that reveals the associated false reality in the 1999 film *The Matrix* (directed by Lana and Lilly Wachowski) as part of their refusal of women's interests. They employ the term "red pill" to identify men's shifted awareness and the associated belief that women are in control. MrSmith, whose member name references a software agent in *The Matrix*, adds a proviso to HotdogGiambi's commentary, noting, "But please continue to shave those legs and armpits ;)."[4] MrSmith's gendered stipulation, which he then nuances by using an emoticon to convey humorous intentions and to rebut critical intervention, supports cultural expectations that women should maintain their bodies and focus on the interests of other people. Women are in a difficult bind because time-intensive forms of self-maintenance are expected but also condemned, and lead to inquiries about women's ability to employ contemporary technologies and be agentive subjects.

Some women's assessments, including the writing of the journalist Michelle Quinn and the online commenters that I study later in this chapter, indicate that iPhone designers do not accommodate women and a range of other people's hands and fingernails. Women commenters identify such refusals to acknowledge women's embodiment as misogynistic. These women's comments are met with sexist retorts. Individuals' responses insist that women and their fingernails are the issue. I argue that this and other occasions where people undermine women's critiques should be attended to because these individuals work to perpetuate hierarchical gender systems *and* because their reactions are often out of proportion to the original reports and comments. As I suggest later in this chapter, such excessive reactions emphasize the workings of feelings in relation to digital technologies and individuals' desires to protect and maintain their brand. In sustaining the technological status quo, these individuals and practices elide the requirements of people who are not considered in technology designs. This effacement of some women, disabled individuals, and their hands occurs through device designs and representations. These people's hands are thus hindered, as computers and Internet sites promote the hand-pointer as a common representation of individuals and screen positionality. While the hand-pointer conveys a representation of individuals' hands without fingernails, popular writing about touchscreens often focuses on the actions and limitations of gendered hands and nails.

In this chapter, I consider the ways individuals manage critiques of touchscreens to maintain their favored brands and normative and agentive male positions. I focus on how women's critiques of gendered designs and scripts, including the ways iPhones privilege men, are understood and dismissed. I engage scholars' research on gender scripts and define the term as the ways technology producers and designers' worldviews influence the production of devices and the associated presumptions about the people who use them.[5] Gender scripts tend to associate advanced technologies and digital aptitudes with men. While the literature on gender scripts ordinarily relates these narratives to designers and devices, I develop this area of study by pointing to the ways individuals who employ and comment on iPhones espouse similar narratives. I relate people's production of gender scripts to their brand identities and the ways individuals are prompted to contribute to companies' positions. Thus, I elaborate upon the arguments of related science and technology scholars about how users matter, which indicate that everyday people, as well as inventors, companies, designers, and marketers, determine how technologies function.[6] My expanded study of gender scripts offers important methods of considering journalism, online commenting, and related practices. I analyze the ways consumers articulate the expected gender and other characteristics of people who engage with technologies. Yet the group of (mostly) male journalists and commenters considered in this chapter dispute indications that gender is a factor in the accessibility of technologies. In doing this, these journalists and commenters establish men's rights through technology scripts and protect their positions. They thereby defend the gender scripts associated with iPhones. They also establish themselves as brand experts and community members.

I provide a study of early reporting and feelings about iPhones. I begin with an outline of how iPhones are correlated with feelings, including the ways Apple depicts people's hands and fingers as key to the interface. I focus on Michelle Quinn's *Los Angeles Times* articles about iPhones and fingernails and the reporting about these texts in numerous newspapers and online forums, including the associated comments.[7] By considering these texts together, I am able to more fully study the common dialogue about and dismissal of women's embodiment in relation to iPhones and the broader notions of touchscreens. In making connections between journalists' articles, commenters' replies, and people's dialogues in forums, I also argue for scholars and general readers to attend to the ways normative gender scripts and the associated worldviews are collaboratively produced. My close reading practices, which consider the varied contributors who engage in and amplify the dispute over the iPhone design, can also be employed when considering other instances of online contention.

When women critique designs and interrogate the associated norms, they and their fingernails are identified as difficult. I employ Sara Ahmed's writing on the willful girl, raised arm, and feminist killjoy to consider how women are dismissed when they conceptually and physically raise their hands and nails and refuse gender norms.[8] Ahmed argues that women become a problem when they point to a problem. My references to Ahmed's and Sarah Sobieraj and Jeffrey M. Berry's work also help me to outline how people link physically touching and emotionally feeling, including their intense reactions to the chronicled debate about touchscreen functionality.[9] Ahmed and Sobieraj and Berry indicate how outrage and other feelings are used to represent and dismiss women. In combination, these texts suggest how women's arguments are distorted and simplified as methods of scripting them as straw feminists.

Popular authors render their own straw figurations by often suggesting that mobile phones are tools that equally empower everyone. Yet their accounts also produce narratives about specific bodies and identities, including indications about gender, race, sexuality, and class. I consider the ways women's fingernails function as cultural signs and gendered digital tools, especially in relation to their utility as styluses for resistive touchscreens, which respond to pressure. I thus rethink the figuration of fingernails, particularly women's nails, as problems. In this chapter and chapter 4, I argue that women's fingernails can be identified as part of their prosthetic handiness and agency. The features of women's fingernails also provide me with opportunities to negotiate the connection between embodied digits and digital devices. The term "digital," as Jack Bratich notes, references both fingers and the delivery of information.[10] The term enmeshes hands and computer and Internet technologies. Mobile phones highlight these connections, including in languages where they are identified as "hand phones," as the disability scholar Gerard Goggin notes.[11] As the figuration of the hand phone suggests, Internet technologies, sites, and online engagements proffer conflicting extensions. These devices and sites extend the beliefs of designers, corporations, and normative individuals and promise to prosthetically combine with and validate their characteristics. The word "extension" also describes artificial nails and how women employ their longer length as a form of reach and a means of rethinking engagements.

Gender Scripts

Nelly Oudshoorn, Els Rommes, and Marcelle Stienstra study the ways technologies convey gendered narratives and notions of users.[12] Their research extends and rethinks Steve Woolgar's analysis of how computers configure the user through instructions, warning labels, and other elements.[13] Oudshoorn, Rommes, and Stienstra call for genuinely configuring the user as everybody rather than only recognizing a specific gendered subject. Companies also tend to claim that everyone is addressed, while supporting only some individuals. These researchers recognize the challenges in facilitating such inclusive addresses. They note that acknowledging everyone matters because scripts embed portrayals of the individuals using technologies and methods of use into the associated objects. Designers' scripts "attribute and delegate specific competencies, actions, and responsibilities to users and technological artifacts."[14] Through designers' scripts, technologies articulate the skill level, knowledge, embodied features, social positions, and practices of imagined participants. Oudshoorn, Rommes, and Stienstra also indicate that when the representations incorporated into artifacts do not match all of the engaged individuals, the associated technologies are likely to fail. Such technological failures, as I suggest later in this chapter, may be managed by suggesting that instances of malfunctioning digital objects occur only with less desirable consumers and practitioners, including women. To be recognized and accepted, individuals have to enact at least part of the identities scripted by designers. In the case of capacitive phones, individuals must be able to easily touch the screen with the pads of their fingers and move their hands and wrists in designated ways in order to be recognized.

Madeleine Akrich is often credited with developing the concept of the script, although earlier psychological literature on script theory considers affect, personality structures, and how individuals employ a form of playwriting to understand relationships.[15] Akrich's theory focuses on the ways designers convey gender scripts through objects rather than how scripts emerge through conversations and engagements between people and social norms. She indicates that designers and innovators inscribe their vision of the world into objects and articulate individuals with "specific tastes, competences, motives, aspirations, political prejudices, and the rest, and they assume that morality, technology, science, and economy will evolve in particular ways."[16] Akrich relates designers' inscription of technical objects with gender scripts to film scripts. In a related manner, Majken Kirkegaard Rasmussen and Marianne Graves Petersen's design research describes the script as a manual that is informed by assumptions about how individuals will use items.[17] Designers' processes of developing gendered technological objects shape how things are used and thereby define the agency of men, women, and other subjects.

Rommes, van Oost, and Oudshoorn identify the pressing problems with gender scripts, since these structures indicate how people should be culturally understood and can societally and technologically engage. They also argue that the "concepts of 'user representation' and 'gender script'" are useful means of studying the "extent to which the problem of under-representation of women as internet users can be understood as a mismatch between the designers' image of users and the actual users."[18] I develop these ideas of gender scripts as a method of considering iPhones and the ways companies and individuals engage with, support, and undermine gender norms. Gender scripts are incorporated into mobile phones and conveyed through a variety of physical and ideological features. In the texts that I examine, people refuse women's critiques of the iPhone design by asserting deeply normative notions of gender and technology use and refuting indications that technologies convey gender scripts. Yet Els Rommes argues that the analysis of gender scripts is a "study of who has to 'adjust' more, who has to pay the price for not fitting the norm that is reproduced in the arte-fact."[19] In addition, as I argue, gender script research should be an analysis of how disenfranchised individuals' experiences and critiques are culturally managed. In the case of iPhones, men and other normative identity-seeking subjects benefit and retain their relationship to the brand by negating women's critiques. Individuals risk penalties for refusing technological mandates and critiquing cultural dismissals, and not all individuals are able to rebut these structures. However, feminist killjoys and willful girls and women, as Ahmed articulates these positions and I develop these ideas in this chapter, offer feminist methods of refusing social control and advocating for a critical feminist life.

The Willful Subject and the Raised Arm

Ahmed formulates the willful subject through a resistant reading of the Brothers Grimm's "The Willful Child."[20] In the story, God allows a girl to become ill because she refuses to listen to her mother. After the girl dies and is buried, her willfulness continues, and she continually pushes her arm out of the earth and raises it up. It is only after the mother comes to the gravesite and beats the girl's arm with a rod that the child withdraws. As Ahmed notes, the story acts as a teaching tool that conveys how girls should obey. She also indicates how the bad, willful girl is culturally distinguished from the strong-willed but acceptable boy. Ahmed writes that the "girl who is deemed willful (from the point of view offered by the fable) is going out on her own limb; she separates herself from her family, an act of separation that is sustained by the transfer of willfulness to her arm, which appears as a limb on its own."[21] Despite the child being beaten down, "some thing, some spark, some kind of energy persists" and the "arm gives flesh to this persistence."[22] Arms, hands, and fingernails, as I suggest throughout this book, can thus be theorized as representations and critiques of humanness and as part of an obstinate critical project.

Ahmed indicates that chronicles of girls and women's aloneness are attempts to undermine the connections and power in linking arms. Women who possess and are possessed by their arms threaten men's expectations of being served. These women may refuse to do the housework and maintain "his house," as Ahmed notes.[23] They thereby decline to engage in bodily labor for others, to be identified as passive feminine bodies, and to support his thinking and position as a male and masculine mind. Women who refuse to be scripted as "helping" and "right" hands are willful subjects who maintain their own agency. Individuals' arms personally and politically matter because they enable people to "reach, to carry, to hold, to complete certain kinds of tasks." Yet laborers lose their arms when employers make them into tools. As Ahmed argues, to "become his arm is to lose your arm" and to free the arms of the factory owner. The act of striking against such forces "is to clench your fist, to refuse to be handy." Such fists have become rallying cries and signs of the revolutionary labor movement and the black power movement. Representations of raised arms and fists are associated with #BlackLivesMatter and #MeToo hashtags and related activist interventions. The combination of the clenched fist and the Venus symbol is a women's movement icon that underscores the power of women and their arms and hands. These images of hands thus can be traced through varied forms of online activism and studied by applying Ahmed's critical reading.

Ahmed considers how willful girls and women are identified as feminist killjoys and are made into simplified and straw feminists. In Ahmed's reparative and positive reading, feminist killjoys indicate instances of sexism, racism, and other forms of disenfranchisement and thereby disturb the ordinary flows of privileged and easy behavior. Feminist killjoys are thus culturally understood as generating and being problems. They are blamed for destroying people's happiness. While feminists are usually identified as getting in the way, feminist technology killjoys are also characterized as intervening in progress. Ahmed notes that feminists are useful as "containers of incivility and discord" and are thought to ask for and to deserve dismissal.[24] Feminist feelings about unjust practices and individuals' reactions to feminist speech also result in sensations. As Ahmed argues, "If sensation brings us to feminism, to become a feminist is to cause a sensation."[25] Feminists and other willful subjects are contained by indications that their resistance to norms is overly emotional, poorly thought out, unreasonable, ill informed, and selfish. The associated willful subjects are further dismissed, as are people who fall outside the usual gender scripts, by rendering them as unskilled, stupid, exterior to cultural standards, and representing beliefs that cannot be maintained. I employ Ahmed's narrative about the willful arm and theorize with it as a way of thinking about the link between normative notions of mobile phone use and individuals who speak up about design conventions and refuse to change their bodily configurations.

The individuals who register concerns about mobile phone designs, and many other women who conceptually raise their arms in protest online, are made into straw feminists and killjoys. In the popular texts that I study in this chapter, the straw person, straw feminist, and simplified versions of the feminist killjoy are often constructed as methods of refusing interrogations of gender scripts. Straw feminist arguments are a kind of straw formulation. According to Jennifer Schumann, Sandrine Zufferey, and Steve Oswald, straw arguments distort an "opponent's view in order to make it more extreme and therefore less acceptable, thus easier to attack."[26] Robert Talisse and Scott F. Aikin's research on the functions of the straw man indicates that people render this figure to misrepresent an adversary's opinions and then contest the distortion rather than the adversary's actual viewpoint.[27] Straw fallacies are also enacted by debating the weakest points of arguments, by incorrectly conveying research and beliefs, and by relying on people's lack of knowledge to advance improbable positions. Such analyses and identifications of "straw man" arguments might be better described as "straw figures," given how often these constructs are used to muffle, denigrate, and displace women, and more specifically feminist women. Straw figurations, such as the effaced targeting of women and feminists, can also be disrupted through close reading and critical theory, including Barbara Johnson's encouragement to attend to ambiguous phrases and words, inconsistencies between what texts communicate and what they enact, texts that protest too much, and irreconcilabilities between contentions and the associated examples.[28]

The columnist Ellen Goodman foregrounds the ways straw feminist portrayals are used to discredit feminists and are "handy for scaring supporters away."[29] Straw feminist formulations are thus designed to control the feelings of people resistant to and interested in feminisms. They are "handy" and accessible methods of denying women's hands, as I suggest in this chapter. Monica Dux and Zora Simic's feminist text evokes the functions of the willful girl's battered arm and straw misrepresentations. They describe how individuals set up a straw "caricature of feminism, built on half-truths, oversimplifications, generalisations and stereotypes," and then "beat the crap out of it."[30] These manipulations of feminist positions are heightened by the methods through which individuals generate outrage. For instance, Sobieraj and Berry explain how people produce emotional reactions by sharing distorted and incorrect information about adversaries.[31] More specifically, methods of dismissing opponents and amplifying allies' indignation include "Name Calling," "Belittling," "Mockery," "Character Assassination," "Misrepresentative Exaggeration," "Ideologically Extremizing Language," "Emotional Display," "Emotional Language," "Slippery Slope" claims, and "Obscene Language."[32] While not always successful, such practices are intended to generate shared anger among instigators and their cohorts. These formulations are also designed to trigger feelings of embarrassment and isolation among those who are disparaged. In the online cases that I study, this generation of resentment and attempts to remove challengers from cultural legibility include rendering the purportedly offensive subjects as inhuman, depicting their bodies as inadequate, indicating that they are intellectually inferior, and suggesting that they are developmentally insufficient and do not recognize the progressive advantages of technologies. Many of the journalists and

forum commenters that I study employ these practices when responding to
women's technology use, including women's concerns that iPhones did not
and do not address their physiognomies and interests.

Apple's iPhone

People's emotional experiences, including excitement and irritation, have
accompanied the release and employment of iPhones and related mobile
technologies, as I suggest in more detail later in this chapter. Steve Jobs
introduced the first iPhone in a keynote address on 9 January 2007 at the
Macworld Conference & Expo in San Francisco. Apple then released the
first-generation iPhone in the United States on 29 June 2007. It featured a
touch-based user interface. Only four male writers—Ed Baig of *USA Today*,
Steven Levy of *Newsweek*, Walter S. Mossberg of the *Wall Street Journal*, and
David Pogue of the *New York Times*—were given review models of the original iPhone.[33] This helped to amplify the desirability of the device that was
initially inaccessible and to associate it with empowered men. Pogue conveys the related cultural anticipation when noting that the mobile phone
was the subject of "11,000 print articles" by 27 June 2007, even though it
had not yet been released.[34] As this suggests, Apple products have been of
great cultural interest, and the features of the iPhone and its impact have
been widely considered in the popular press.

A large number of reports about iPhones consider the touch interface,
including the ways people text and input other information by employing the simulated and screen-based keyboard. According to Pogue's early
review, "Tapping the skinny little virtual keys on the screen is frustrating,
especially at first." He thus connects physically touching the screen, the
somewhat ephemeral features of the represented keys, and feeling irritated.
Walter S. Mossberg and Katherine Boehret's review indicates that generally
the keyboard "works," but "you have to switch to a different keyboard view
to insert a period or comma, which is annoying."[35] Mossberg has more
recently suggested that during his initial three days of use, he "was ready to
throw this thing out of the window for trying to type on glass."[36] The limitations of shifting between different keyboard configurations and persistent
typing errors are also issues with contemporary versions. In 2017, about 50
percent of the emails Levy was receiving still indicated, "Typed on phone,
forgive typos." Thus, the design of screen-based keyboards results in typos

and a culture where such limitations may be excused. Certainly, individuals' varied forms of online communication, including their synchronous text-based chats, acknowledge and grapple with typing errors. People's vernacular online texts incorporate references to teh (the) and pron (porn) and thereby acknowledge the limitations of interfaces, the inabilities of members, attempts to circumvent filters, and particular interests. Quotations of online texts with typos represent the ways fingers work (and do not work) on touchscreen and other keyboards.

Many early reviewers of the iPhone underscored the limitations of its keyboard interface. However, Jobs asserted in his Macworld keynote that the "iPhone does not use a keyboard, nor does it use a stylus, as many smartphones do."[37] He thus differentiates between screen and button inputs. He continues to distinguish the iPhone by arguing that the device features the "best pointing device in our world." He elaborates, "We're born with 10 of them, our fingers." Jobs thus figures the iPhone and ten-fingered embodiment as natural and an inherent part of an intuitive interface. Embodied digits are represented as tools for phones and connected to digital devices. Through such notions, Jobs and the designers of the phone produce a script that asserts what constitutes a body and how the body functions. They configure the user by providing a framework and instructions on how to engage. These instructions may already cause misdirection because few people use all of their fingers with touchscreens. In 2010, iPhones still often only recognized three fingers being used simultaneously. The different ways individuals employ phones are compressed and exaggerated by the idea that iPhones are best, and that they are best activated and experienced with ten fingers. Yet constituting this everybody, which is actually a very specific kind of body, obscures different embodied configurations and life stories.

Jobs's claim that fingers are the best pointing device is challenged by the many people who have sought iPhone styluses and by Apple's sale of an Apple Pencil. This pencil was marketed along with the iPad Pro in November 2015, and a second-generation pencil was promoted in November 2018. Apple describes the second-generation pencil as setting the "standard for how drawing, note-taking, and marking up documents should feel – intuitive, precise, and magical."[38] Apple thus continues to suggest that individuals will have a "magical," effortless, and unmediated experience with its products. According to the company, the pencil is a "comfortable, natural tool" that has a "seamless design" and can be "engraved to really make it your own." Apple renders its varied styluses and interfaces so that they negate the technology in favor of natural and personal experiences. The associated advertisements depict a dark-skinned hand rather than Apple's ordinary emphasis on light-skinned proficiency (figure 1.1). However, the very white pencil conceptually displaces this hand. The pencil is demarcated in this instance as the "ideal tool" and more digital than the hand. The dark-skinned hand may itself be marked as configurable and already replaced because a color palette below the hand provides an array of possible skin tones to choose from and shows the lightest flesh color selected. These representations of configurable bodies, and the associated need for close readings and other scholarly interrogations, also appear on clothing, exercise, and gaming sites, including the marketing of hand- and body-extending peripherals.

Jobs's emphasis on fingers is echoed in Quinn's 2007 *Los Angeles Times* article about Apple's initial iPhone release. She reports, "Jobs pinched his fingers together on the screen to shrink a photo, then spread them to expand it."[39] Fingers are further foregrounded because the article is titled, "Latest technology is at your fingertips." Quinn mentions fingertips as a

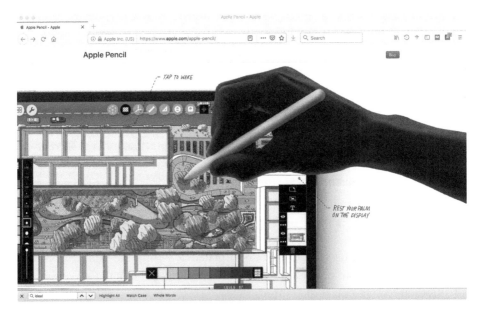

Figure 1.1
Screenshot from the "Apple Pencil" advertisement.

means of conveying availability and access, as well as individuals' processes of activating touchscreens with their hands and fingers. She distinguishes between stylus and touch-based phones. Quinn also cites the Apple representative Natalie Kerris's dismissal of the "penlike pointer as 'inconvenient' and 'easy to lose.'"[40] Kerris echoes Jobs's claims when asserting that it is "more natural to use the pointing device we were all born with -- your finger." Kerris supports the embodied script established by Apple, and conveyed by Jobs, and helps to configure the normal body and the ways it works. Such able-bodied figurations echo the structures of Apple's interface in suggesting that there is no mediation between bodies and the delivered information. Apple's varied references to finger-pointers and their functions as part of the digital interface evoke hand-pointer icons, which appear at varied junctures when employing computer and Internet programs and interfaces. The hand functions as an embodied, branded, digital, and representational device to connect different Apple experiences and products. While Jobs and Kerris claim that fingers provide the best pointing device, reviewers like Pogue have resisted this finger (and hand) pointer and identified typing as "tapping" and exasperating.[41]

Quinn and the "iPhone fingernail problem"

Quinn's 2008 article, which is titled "Finally, Steve Jobs unveils iPhone 2.0 and iPhone 3G," conveys anticipation for and positive cultural sentiments about iPhones.[42] The title and Apple's sequential numbering frame the device as a development and an upgrade, which, as I suggest later in this chapter, works with other narratives to render women with nails as undeveloped. In the comments section of Quinn's article, Erica asks why designers persist in ignoring the needs of people with fingernails.[43] Her comment indicates that the promises of upgrades and advancements have not been delivered. EWC, who Quinn cites in a later article as Erica Watson-Currie (and as a different commenter than Erica), declares that she wants an "iPhone with Stylus!!!"[44] Her "husband LOVES his iPhone, but with fingernails," she finds it "frustratingly difficult to use the keypad." She adds that there are instances where her nails function as useful tools. With her iPAQ, she does not even have to employ the "stylus: just tap with fingernail & it's gold!" However, her relationship to Apple is one in which she "waited - patiently, G*D**IT! - wishing & hoping that Apple in it's infinite goodness

& wisdom would just open up a stylus option" but "NO LUCK! :(." She waits for and desires recognition and a device that is better configured for everyone. Watson-Currie's interrogation of Apple's lack of recognition should also be applied to other technologies and to culture more broadly.

Watson-Currie's critique is in response to a situation where the gender script of iPhones fails to recognize certain individuals, and where the design actively stymies the interests and skills of women, other people with fingernails, and individuals with other embodied differences. She references failed brand attachments as a means of trying to get readers to emotionally engage in and support her critique. My close reading of this and other texts, including my study of hearting in chapter 3, is informed by the ways passionate expressions like "LOVES" and "G*D**IT!" and the employment of uppercase and punctuation for emphasis are intended to amplify the argument and associated sentiments. People's use of all uppercase letters online is often identified as shouting. As Your Uncle Cordelia dismissively suggests on Urban Dictionary, uppercase is rejected because it is thought to be performed by "tweenie girls, newbies and the mentally unstable."[45] People's employment of uppercase thus functions as a form of Ahmed's raised arm, which other individuals dismiss by associating uppercase communication with identity stereotypes.[46] They reject persistent capitalization, and the emphasized visibility of the associated people as means of discouraging different expressions and sentiments.

Quinn references sentiments when describing the "much-anticipated iPhone" and the associated finger gestures. She then provides an intervention into Jobs's, Kerris's, and Apple's establishment of the finger-pointer as inherently human and intuitive. Quinn notes that "for many years, researchers couldn't master the finger." While she probably intends to suggest that researchers could not design adequate touchscreen interfaces that worked with finger input, her employment of an ambiguous phrase, which Johnson foregrounds in her close reading practices, positions researchers as uncoordinated and unskilled technology users.[47] Quinn's statement is the same as noting that researchers could not master the digital. It also suggests that the company could not emotionally deal with consumers who responded by giving it the middle finger when grappling with difficult devices. These researchers are repositioned and situated at the level of the body and rendered as unable to coordinate emotional, embodied, and technological features.

Quinn adds that researchers' "efforts to create sophisticated touch-screen computers were beset with technical challenges, high costs and doubts over whether consumers, trained to treat the mouse and keyboard as extensions of their hands, would adopt touch screens."[48] She provides an invention narrative about designers' persistence. Of course, chronicles of the finger-pointer are also accounts about technological invention because they make hands into extensions of keyboards and devices. While perhaps clarifying her indication about mastering fingers, Quinn's comments still trouble normative notions of how individuals employ technologies and how prosthetic bodies are configured. Trained and configured consumers are presumed to envision their bodies as extended by mice and keyboards, and thus to refuse touchscreen-only inputs. Her comments may also indicate that such peripherals as the mouse are understood as extensions of the body, while touchscreens that work with finger and hand inputs efface the screen (and perhaps the hand) in favor of an unnoticed interaction.

Quinn's related article from 2008 is titled "The iPhone fingernail problem."[49] She begins by arguing that there has been a cultural shift in women's empowerment, including the ways "Hillary Clinton broke new ground in her race for the White House" in 2008. Quinn then suggests that some women identify iPhones as exceptions to this advancement. Her narrative about a recent cultural shift in women's rights should be interrogated in relation to varied forms of misogyny, LGBTQIA+ phobias, racial intolerance, and the delimitation of voting rights and sexual health options. This includes Donald Trump's dismissals of Clinton and racist denigrations of women of color as methods of consolidating his power and speaking to his male and white supremacist base. Quinn's reporting takes up similar concerns when noting that "some iPhone users complain that when it comes to the hot gadget from Apple, women are still being treated like second-class citizens." She recognizes the critique, but her word choice also risks suggesting that women are selfish complainers who do not recognize their own advancements. She references Watson-Currie and other individuals who posted critiques of the iPhone in the comment section of Quinn's earlier article. Quinn chronicles Watson-Currie's frustration because the iPhone's improvements have not resolved the "fingernail problem" and have not included a stylus. Watson-Currie articulates a cogent critique when asking, "why does Apple persist in this misogyny?" Quinn quotes this text in her article about the "fingernail problem." However, Quinn also offers a broader account of social change that flattens such analysis.

In Quinn's and associated journalists' iPhone reporting, women become a problem. Such articulations of women and femininity as problems have been ongoing political and cultural figurations. In his writing about femininity, Sigmund Freud notes that to "those of you who are women this will not apply—you are yourselves the problem."[50] Mary Ann Doane argues that Freud "evicted the woman from his lecture on femininity" because she is too "close to herself, entangled in her own enigma, she could not step back, could not achieve the necessary distance of a second look."[51] She uses women's performances of femininity and overt closeness to their images, as I detail in other research, to argue for critical forms of masquerade and ways of critiquing women's association with objects and representations.[52] In writings about iPhones, women are at once too close to themselves (and purportedly enraptured by their fingernails) and too distant from the device, contact, and screens because their fingernails hold them at a distance. Or, at least, women are imagined at more of a distance to the device than men's bodies, which are presumed to be "without" nails. While women may be correlated with physically touching and feeling, in these cases men are figured as being better able to touch the device and therefore empowered. Of course, this rendition of closeness to the screen and device is different than the usual desirable distance from cinematic screens and other representations, which have been associated with normative men's masculinity.

Screens are often associated with control and power, including buttressing the purportedly ideal position of white heterosexual men. However, Quinn's title may figure the iPhone as embodied and as having a problem with its fingernails. The title also renders the fingernail as broken and a failure. Yet her article is identified as a problem by many commenters, as I indicate in detail later in this chapter, because it reports on women's concerns about phone affordances. In addition to the use of the term "problem" in the title, Quinn employs repetition to facilitate the kinds of emotional intensification that I discuss in the introduction, and she uses the word nine times in her article. Many of these instances specify the "fingernail problem." For instance, Quinn uses the term "problem" when conveying evaluative distinctions between functionality and aesthetics. She writes, "Problem is, the iPhone's touch screen responds to the electrical charge emitted by fingertips. And pretty though they may be, fingernails don't emit one." Quinn thus figures fingernails, and potentially women with fingernails, as aesthetic objects that have no functional purpose. Of course, most people have fingernails, so these distinctions become murky and often are associated with gender conventions rather than nail length. Quinn and others have argued that women should buy styluses and, more dismissively, that they should cut their nails. Yet Watson-Currie evokes the literature on gender scripts and designing for everybody when she refuses suggestions that women should trim their fingernails. Watson-Currie argues, "It's the machine's job to accommodate its users, not the other way around."[53] Of course, the machine has to be inclusively designed and manufactured. Erica (who shares a first name with Watson-Currie but Quinn identifies as "another woman") asks in Quinn's previous article, "Why are they still discriminating against those of us with fingernails?"[54] As these commenters suggest, women with fingernails are expected to adjust their bodies to the technology because they do not meet the norms articulated by the company, designers, and device.

Quinn indicates that women "complain," rather than acknowledging that their considerations are vital critiques. Women's positions are denigrated through the cultural identification of women as complainers and their critiques of Apple designs as complaints. As Ahmed argues, a "complaint brings you up against the culture of an institution; and a complaint is often necessary because of the culture of the institution."[55] Women and their critiques are dismissed through the emotional practices of incitement that Sobieraj and Berry describe, including belittling women's choices and misrepresenting their concerns.[56] Women's critiques are turned into complaints as a means of dismissing their queries as unnecessary and as slippery-slope arguments that will ruin people's attachments to the company and its devices. Identifying women's analyses as unreasonable objections allows people to remove women from the category of ideal device users. These straw figurations advance the forms of gender management that justify the treatment of women as difficult and their technology use as suspect. Ahmed indicates that the "experience of a situation as something to be complained about is an experience of coming apart from a group." Women are informed that they could have been part of the technology and Apple community if they just knew how to engage with devices and be mannerly citizens, although this inclusion is always provisional and with limits, and women are told that it is their fault that they are no longer embraced.

Quinn updated her report with "some of the criticism" that the "coverage of this fingernail problem has generated." She writes, "Radio disc jockeys have howled in derision, as have commenters and bloggers, about Erica Watson-Currie's critique that Apple was being unfriendly to women and 'misogynistic.'"[57] Quinn references and may further amplify the kinds of production of outrage that I have previously outlined. She undercuts considerations of how technologies like the iPhone perpetuate gender scripts by putting the term "misogynistic" in quotes and thereby appearing to use scare quotes and contest women's critiques. Quinn's highlighting of the term "misogynistic" also appears in the criticism that she mentions. She further promotes the common dismissive response: "They said: Cut your nails! No one is making you buy the phone!" Such comments indicate that people with longer nails, who are often women, have no rights and claims to group membership and citizenship. They are expected to make accommodations for or else not engage with capacitive mobile cultures. These texts begin the supremacist process of distinguishing between people who are understood as being fully human and having hands, as I outline these cultural distinctions in the introduction, and women and other dismissed subjects who are believed to be less-human creatures with nails and claws. While women and other subjects with fingernails are disparaged, many men, as indicated in the reports and comments about Quinn's article, are comfortable with the iPhone interface and the related gender scripts.

Women cannot meet gendered iPhone scripts when they follow cultural expectations about grooming and/or embrace a number of cosmetic practices. Sherifftruman foregrounds expectations about feminine beautification and self-maintenance in a MacRumors thread about "Women's Nails vs. iPhone." His wife works in management in a "pretty large company so long nails are out but having nails that look good is pretty much a must when you meet clients."[58] While she "texts fine and emails some" on her iPhone, "she'll never be quite as fast" as with her Blackberry. Thus, as suggested by Sherifftruman, iPhones are not designed to recognize and do not work well with the kinds of fingernails that women are culturally expected to maintain. This framework allows Sherifftruman to distinguish between women's managerial nails and longer and unacceptable lengths. In these instances, women are forced to be a corporate problem because of cultural expectations about grooming, or to experience and identify the phone as a problem.

Sheriffruman employs class distinctions when distinguishing between acceptable managerial fingernails and nails that are too long. Quinn promotes related gender and hierarchical scripts when describing the "age-old struggle between tech lovers and regular people over how user-friendly gadgets and computers should be." Quinn's equation renders all women with fingernails as nothing more than end users, who are distinguished from inventors and the technologically skilled. Of course, women can be tech-lovers and technological experts and still have difficulties using the device because of the ways it is designed. Science and technology studies scholars have noted how distinctions between designers and technology users tend to support the binary of skilled and active men and passive women consumers.[59] Such scholars intervene in distinctions between inventors and users by underscoring the ways individuals determine the uses of technologies.

Quinn addresses these technologically produced end users and "those who choose their nails over the iPhone."[60] She warns them: "Discussing the unfairness of being forced to choose between the two could invite ridicule." As she chronicles, there are numerous negative online responses. Many people's online comments dismiss Quinn's evaluation of the interface, the reporter, Watson-Currie's and other women's critiques of the device and company, and the feminist language of demanding recognition and rights that these women's interrogations call upon. Quinn's warning is mainly directed at women and encourages them to accept design norms or experience the kinds of incivility outlined by Sobieraj and Berry. Quinn's article also includes a representation of Watson-Currie's extended arm (figure 1.2).[61] She holds a phone and uses her fingernail as a stylus. This arm and nail, which function as a protest, echo Ahmed's willful subject. However, Quinn's text acts as a version of the mother's rod and a warning about the cultural sanctions that batter arms that dare to resist. A version of the image, which is titled "Woman with fingernails attempting to type on an iPhone," appears in her subsequent article and helps to push down women's protesting arms and abilities.[62] Quinn's warning about being mocked is buttressed by the numerous negative reports and comments about Quinn's articles and Watson-Currie's commentaries. Journalists' and commenters' adverse indications, which I consider in the following sections, are designed to convey that women who question technological affordances and gender scripts will be mocked and repressed. Responses to the articles and associated

Figure 1.2
Screenshot from "The iPhone fingernail problem," with Erica Watson-Currie's photograph.

women, which are out of proportion to the number or influence of the texts, reinforce the brand affinities and gendered terms of technology use.

Commenting on a "fingernail problem" in the *Los Angeles Times*

Commenters take advantage of the title of Quinn's article and her constitution of a "fingernail problem" as a means of reinforcing behaviors and norms. BB user writes in the associated *Los Angeles Times* news forum, "that's YOUR problem and not Apple's."[63] BB user suggests that the critics and critiques, as well as their fingernails, are the issue. BB user disputes women's interrogations of the ways technology companies and designers fail to acknowledge different subjects and forms of embodiment. According to the commenter, "They don't have to design it for everyone, and just because some women CHOOSE to have long nails and can't use it does not make Apple misogynistic. A man with long nails would have the same problem. For the fat-fingered, same thing, buy something else and quit your griping!" The commenter's indications unintentionally underscore that iPhone designs fail in numerous cases. The notation that other subjects remain unaddressed by iPhones does not support the idea that such failures are acceptable and does not demonstrate that Apple and other technology companies script their products with all gendered consumers in mind.

My close reading of BB user's commentary indicates the methods that individuals use to support the existing gender order, including making false equivalences, providing other examples as a means of suggesting that it is women who are discriminating, and constituting an "almost everyone" that devices satisfy and a purportedly small cohort that need not and cannot be supported by designers and products. In addition, such texts constitute women as whiners, render a structure where women are denigrated for "griping" and "complaining" when they express concerns, and tell women to shut up. Yet BB user employs capitalization that conveys shouting and has been dismissed online as being a problem. As Ahmed's analysis of the feminist killjoy suggests, Quinn, and especially Erica and Watson-Currie, point to a technological and sexist problem and are turned into the problem. These women illustrate instances of misogyny that people do not want to acknowledge. Individuals try to displace the critiques and associated design issues by portraying the women as annoying and at fault, even as supporters of the iPhone also engage in behaviors that have been culturally dismissed.

Women are constituted as the problem in the same forum when Dan M. writes, "Get a grip lady. Another example of spoiled American women not EVEN knowing how good they have it. Ever heard of Pakistan?"[64] These commenters try to dispel Watson-Currie's and other women's critiques by distinguishing between commenters' virtuousness and women's purported lack of concern about other forms of disenfranchisement. Watson-Currie and the women she is deemed to represent are portrayed as unknowledgeable about a subject that is not related to the article or women's observations. This is an attempt to misdirect the debate with straw positions and to render the related arguments as selfish and ill informed. Women are depicted as ungrateful, and in a move that evokes anti-feminist men's groups, as beholden to the men who purportedly labor without recognition. Dan M. also stresses his reasonableness when rejecting the "lady" who is supposed to be lacking a "grip" and mental stability. He deploys the kind of name-calling that Sobieraj and Berry mention. Such dismissals are part of online argumentative methods that are designed to repress discussions and dismiss already disenfranchised people.

Xtopher distorts the argument when inquiring, "Useless for the blind! Yes. And useless for people without any hands! What was iPhone thinking? And the deaf! They can't use it. iPhone should fix that. And the stupid!"[65] In these cases, commenters continue to compare straw versions of the reported critique to what they identify as more important device limitations as methods of rejecting the argument and rendering critics as stupid. Dan M.'s and Xtopher's examples do not reconcile with Watson-Currie's critique, and thus they enact a version of the evasive writing strategies that Johnson suggests that readers study—approaches that are also useful in Internet research.[66] While Xtopher indicates that considering the use of iPhones by people with disabilities is stupid, technology companies have been critiqued for their effacement of individuals with disabilities and been mandated to provide affordances for the disabled.[67] Gerard Goggin correlates touch, disability, and feelings when outlining how early versions of the iPhone caused the blind community to be "enraged" because of limited access options.[68] He also describes more recent VoiceOver additions, virtual controls that use finger gestures, and TouchID, but such changes often require skin-to-phone contact.

Commenters do not emphasize the histories of technology and disability, and instead work to undermine women's interrogations of misogyny. Historical studies of related phenomena include Carolyn Marvin's research on how early phone use by women and people of color was stereotyped as unintelligent and excessive.[69] Online commenters also continue to gender-code women critics, including Xtopher's association of women with stupidity. Such portrayals validate the exclusion of women by depicting them as incapable of technology use. Xtopher and others reference disability as a means of asserting their greater attention to difference and as methods of negating women's critiques of iPhones, but such commenters also employ an ableist approach that uses cognitive disabilities as a way of rejecting women.

Quinn, Watson-Currie, and Erica underscore that iPhones do not recognize women, who are the largest population with longer nails. Nevertheless, George Kaplan claims, "We've arrived at a time when each person expects the world to adapt itself to meet his or her needs."[70] This constitution of the individual selfish woman is also part of the accusations advanced by anti-feminist men's groups, which turn feminist critiques of social systems into the demands of privileged individuals. They thereby try to roll back perceived losses in men's privileges and try to prevent cultural shifts that would provide more rights for disenfranchised groups. Kaplan and related commenters indicate that a technology cannot meet every individual's needs, and such expectations are self-centered and unreasonable. If Apple offered an array of devices that recognized a larger group of individuals, it would not inherently influence commenters' experiences.

Kaplan produces a common argument when writing, "She can cut her damn nails, or she can buy a Blackberry. It is called freedom of choice." In a related manner, the commenter iPhone user indicates, "Don't like it? Go get a Blackberry, don't turn" the commenter's "iPhone into your black-berry."[71] These kinds of neoliberal frameworks shift the responsibility for providing equitable services and meeting people's needs from the state and corporations onto individuals. References to choices may be particularly insidious and enact straw feminist tactics. Choice has been a persistent femi-nist concern, including attempts by second-wave feminists to intervene in the cultural repression of women's reproductive and other choices.[72] Some third-wave feminists endorse individual women's rights to make personal choices. Linda R. Hirshman's research on women and work is associated with the term and critiques "choice feminism."[73] Feminists have also questioned some of the associated choice feminism formulations because they risk sug-gesting that women's decisions are easily made, and they are personal rather than influenced by cultural expectations. It is these tactics that Kaplan and iPhone user narrow and repeat.

Commenters do not consider equal and equitable engagements when they indicate that men's bodies are the standard, and women and other subjects should change to match men's experiences and expectations. As part of this structure, these commenters promote a series of evaluative judgments about mobile devices, including the identification of iPhones as masculine and the property of men, and Blackberry phones as outdated technology, less desirable, and feminine. While there are other indications that iPhones, and phones and their casual communicative functions more generally, are associated with women, the continued dismissal of women's queries about capacitive screens and fingernails suggest the limits of such gender coding.[74] Phones tend to be rendered as feminine and the domain of women when they are marked as social and frivolous.

Clawing at Hands in the *Los Angeles Times*

Informational videos about the technical specifications of devices often render working and technologically proficient hands and suggest that these appendages do things. Apple designers and technologists' hands, which are often depicted outlining technological features in Apple videos, are almost always male and are often white. The 2018 site for Microsoft's Surface Pro depicts the body and hands of entrepreneur Adam Wilson using the device to "bring his ideas to life."[75] His hands are thus associated with human inventions and godlike skills and powers. Hands, as I suggest in the introduction, are often used to articulate the features of humans that purportedly distinguish people from animals. Dr. GLK uses such cultural hierarchies when replying to Quinn's reporting in the *Los Angeles Times* and indicating that "people with claws should not buy iPhones."[76] Dr. GLK's "cats had the same issue, until" he "trimmed their claws." Dr. GLK correlates women with animals, which is a common misogynistic practice that renders women as requiring intervention and being inhuman.

Women are often marked as ugly dogs, controlling and mean bitches, sexual bunnies, and soft and delicate does and kittens. Figuring women's nails as claws is a method of representing the associated women as animalistic, out of control, endangering, and manipulative (when women are said to "get their claws into" people). Narratives about cutting animals' and women's claws are metaphorical and material attempts to diffuse their power. Dr. GLK's comments thus suggest that human subjects can and will retrain women's bodies according to the requirements of devices and what they perceive as normative cultural specifications. These forms of gender patrolling are designed to constrain "complaining" women and to restrict feminine nails that would ordinarily satisfy cultural norms. Women should be able to grow long nails and be acknowledged by Apple's design scripts, especially when these technologies are marketed and identified as innovations that address everyone.

Innovation narratives are employed by emcee as a means of dismissing Quinn and Watson-Currie. The commenter argues that it is a "sad, sad, sad world we live in when people actually have to complain about a high tech gadget that does infinitely more than any computer could 20 years ago."[77] By identifying technologies as forward thinking, emcee and related commenters render women and their critiques as backward. Insinuations that women are devolving and undeveloped also are made when emcee writes, "If you have time to generate a story like this based on the length of your nails you are the lowest common denominator in human evolution. Do humanity a favor and get out of the gene pool." The author thus positions Quinn and her cited authors as insignificant and suggests that they be sterilized, or even euthanized. Oz makes related cognitive distinctions when asking, "Are you that retarded that you need to write a long article about why they should improve the iPhone."[78] These tactics have historically been used by eugenics movements to control nonnormative people and facilitate supremacist worldviews. As the historians Philippa Levine and Alison Bashford note, eugenics is based on an "evaluative logic."[79] Eugenicists employ reproductive practices, claims about medicine and science, and propaganda to assert that the lives of some people are more valuable. Oz and emcee employ sexist and ableist dismissals as methods of rendering women as technologically backward and men as evolved and appreciative of mobile devices. emcee offers a slippery-slope argument where women's critical interrogations threaten cultural and biological developments, and thereby evokes supremacist interests in white male dominance. These commenters dismiss Quinn's coverage of the usability of iPhones, but it is her job to consider these issues.

RRaya also references progress when writing that "this is the future and back in the early days it was required for women to have longer finger nails to scratch" their "husbands backs."[80] The comment conveys a kind of temporal distortion in the form of a contemporary future where women need to catch up. According to RRaya, nails are part of a historical past that should be left behind, and women with fingernails are retrograde. This may evoke a humorous tone, but as the writing of Ahmed suggests, women's arms and hands are culturally demarcated to be at the service of men. RRaya states, "those days are long gone so why are women still growing long finger nails?" The poster thereby renders a situation where it is women who are still supporting and offering their arms to men. RRaya also makes distinctions between having nails and performing "real" dirty work, a false distinction that I critique in other research and later in this chapter, by writing, "Girls cut those nails and start getting those hands dirty the time for change is now." This renders women as regressive, resistant to change, and at fault for their situation. In addition, such individuals presume that cleanliness and grooming cannot be a part of or be related to labor, even though aesthetic work is labor. According to such commenters, women are responsible for damaging cultural practices and should support the associated technological improvements, even as women's attempts to question related scripts are negated.

Reporting on Women in Other Forums

Quinn's article was widely remediated, which resulted in Quinn and Watson-Currie receiving more dismissals. Unsympathetic portrayals are used to justify, while reenacting, the online harassment experienced by women, LGBTQIA+ subjects, and people of color. These connections are foregrounded when Watson-Currie reflects on the writer and media entrepreneur Ellen Brandt's blog post about "Life With a Twitter Stalker."[81] Watson-Currie chronicles how she "diss'd the iPhone" and ended up "sorting the hate mail." She originally thought that the "kerfuffle might result in an article or other media project." However, the "daily barrage" affected her, and she "abandoned the field."[82] The blogger s.e. smith explains in another forum that the point of online threats is to "silence" feminist and other critical voices.[83] Some women, faced by hostile comments and threats, decide to stop researching online hate (as Watson-Currie did), cease posting to their blogs, adopt practices that make their texts less easily located (and their critiques less visible and effective), stop using social media, and cease covering feminist subjects.

Targeted women indicate that the companies that run social media sites and law enforcement and government agencies often provide no support. For example, the reporter Aja Romano describes the all-too-common experience of going to the police after receiving death threats and being ignored. She describes the police being shown violent images, but not appearing to understand that physical assaults can be tied to such online threats.[84] Despite such instances of police and governmental lack of interest, online threats and obsessive attention have resulted in violent actions against women and other oppressed individuals. For instance, a fan of the gamer and YouTube producer Meg Turney came to her house with plans to murder her husband, but he killed himself instead.[85] The gamer Bianca Devins was allegedly murdered by a man she met online.[86] The accused killer posted images of her dead body and livestreamed his attempted suicide.

People use articles like Quinn's to mobilize wide-scale gender patrolling and to maintain the most rigid beliefs and behaviors. Commenters who denounce Quinn's and Watson-Currie's considerations associate the reporting and critique with feminist journalism. They are thereby able to activate anti-feminist and anti-woman sentiment. Paul responds to Quinn's article by asking, "WTF? **Misogynistic?** Who was the woman? Amanda Marcotte?"[87] Marcotte is also referenced as a form of dismissal, and her journalism is mocked in anti-feminist online men's sites, even though her feminist reporting appears in *The Guardian, Salon,* and other mainstream publications.[88] Paul employs this reference to reject the feminist labor of women who interrogate the gender scripts that render technologies and Internet sites as male terrain. Attempts to silence and dislodge the associated feminist killjoys and willful subjects, including comments and threats that are designed to upset feminist bloggers and make their posts seem emotionally fraught, are productive for individuals who harass people online. Feminists, LGBTQIA+ subjects, and people of color might resist and reframe such harassment by reporting on and making intolerant practices visible; supporting individuals who are being doxxed (having their addresses and other personal information published so that they can be located and physically challenged at any time); forming collectives that offer proactive strategies for survival and resistance; informing institutions that such attacks include attempts to discredit scholars and employees; insisting that hateful content about other people is not an acceptable part of online sites; and critiquing the interfaces and beliefs that enable hate.

Online hate and the associated attempts to further oppress people are often fueled by beliefs that advanced Internet and digital technologies are for men. People's insistence on men's ownership of settings are informed by online lore and everyday practices, including the GamerGate targeting of women game designers and critics. The "Rules of the Internet" and its purportedly ironic indication that "there are no girls on the Internet" renders female contributors as childish "girls" and suggests that it is funny to erase their contributions.[89] In a related manner, the journalist Nick Farrell undoes women's connection to technology and justifies their dismissal by titling an article, "Apple fails to tap female market: Iphone is not for girls."[90] Quinn's article articulates a problem, but Farrell works to resolve the association of iPhones with everyone by suggesting that these devices are solely for men. He also reduces women's authority by referring to them as "girls."

While many of these journalists and commenters depict gender critiques as "complaints," Farrell escalates this dismissal by describing Watson-Currie's comments as "moans." Such narratives echo cultural notions that women are too loud, difficult, and excessive. They inform women that they should speak less often and less confidently, and thus follow constraining gender scripts, or be mistreated.

The writer Ken Levine also tries to curtail indications that Apple's design was misogynistic, and women are effaced, by indicating that he "can't imagine the Apple design engineers all getting together and saying, 'Women won't date us for some completely unknown reason. What can we do to get back at them?'"[91] Levine creates a straw scenario that is different than women's indications. Yet women's interrogations of sexist behavior, as I suggest in *Producing Masculinity: The Internet, Gender, and Sexuality*, are often controlled by positioning them as heterosexual and available objects.[92] Levine employs this tactic and resituates women as erotic objects that are to blame for cruelly denying men. This is intensified by including photographs of women in low-cut outfits that appealingly look out at "male" viewers while holding phones. The visual argument therefore also indicates that women's arms, and breasts, should always be focused on men's interests. Levine repeats the terms "misogyny" and "misogynistic"—a tactic that I consider in other close readings—as a method of continuing his straw exaggeration and eviscerating the critique of Apple's gender scripts. While Levine creates the scenario in order to render women and their ideas as preposterous, women are often warned and threatened with punishment for diverging from cultural norms. Women are advised in social and online situations, including writing by anti-feminist online men's groups, that women's expectations that men should treat them as equals and not sexualize them make women undateable and valueless. Many women are also regularly barraged, as smith, Romano, and other feminist journalists and scholars indicate, with rape and other violent threats.[93]

Forum Posts About Gendered and Raced Fingernails

Levine, in a similar manner to posters who mention oppressed subjects as a means of critiquing Watson-Currie, tries to distract readers from her argument. He asks, "Are contact lens manufacturers also misogynists?" and the companies "who made rotary phones? Those couldn't have been easy to dial with Vampira nails." Other critics compare fingernails to the "dragon lady." This stereotype is evoked when alexthechick conveys "little sympathy for those women with dragon lady nails who also whine that this makes practical things like typing and masturbation more difficult."[94] As Shoba Sharad Rajgopal suggests in a critique of representations of Asian women, the dragon lady stereotype depicts Asian women as hypersexual animals who unsheathe their nails, expose their fangs, and growl.[95] Nails situate such women of color as external to the reasonable and human. People's equation of nails to the monster (and vampire) is also a gender script because such creatures are often identified as feminine and outside cultural norms.

FITCamaro tries to establish the similar claim that instead of women "being a stereotypical piece of trash with long fingernails," they should "just get rid of those ridiculous things."[96] FITCamaro's member name invokes gendered interests in automobiles and the associated masculine aesthetics, while dismissing fingernails as feminine and valueless (or negatively appraised as lower class). And kileil concludes, "Throw these women in the garbage."[97] In each of these cases, I attend to and encourage readings of word choices as important parts of analyzing how individuals code identities. Commenters associate women with "trash" and "garbage" as methods of undermining women's value and proposing to eliminate, or "throw away," the associated women. These commenters employ dismissive, even violent, language that rejects women's practices and the associated women by rendering them as tasteless and lower class. They refuse interrogations of how technologies render women as second-class citizens while producing gendered hierarchies and class distinctions.

Commenters chide women for their taste and aesthetic decisions as a means of controlling women's appearance and behavior. For Tim W., fingernails are impediments to women's standing because they are associated with "Hookers, housewives from Long Island with big hair and 14 year old girls who are suddenly allowed to use makeup for the first time."[98] Noticeable fingernails are also conflated with sex work when brian writes, "Long nails are gross" and "make women look like porn stars."[99] brian tells women to cut their nails to a "business length and stop thinking that the world revolves around you." Yet such directives suggest that the world should revolve around men who imagine themselves as judges and arbiters. These male commenters dismiss women for being associated with sex work, while scripting them as sexual objects that are designed for men. For instance,

Sebastian uses humor to produce a gender script when commenting, "If their fingernails are too long for the iPhone then they are also too long to give handjobs."[100] Such commenters perform their normative heterosexuality and try to advance men's power and sexual interests. They thereby continue the cultural structuration of women's bodies, and arms, as designed to address and to satisfy men. Their ideas that women's bodies are configurable and formed for the interests of others influence understandings of women's fingernails.

These posters model the kinds of feelings and indignation that Sobieraj and Berry outline, which lead to misrecognition and miscommunication. This happens when sxr7171 suggests, "She has a lot of nerve to consider that an issue that specifically pertains to women. Always trying to find some way to make every problem on earth a 'woman only' thing."[101] sxr7171 engages in exaggeration when indicating that critics suggest all problems influence only women. Such figurations allow posters to employ outrage as a means of refusing women's demands for more rights and interrogation of the normative gender scripts that situate women as outside of or peripheral to technology use and design. Rather than identifying the bodies that Apple and other manufacturers have recognized and missed, sxr7171 creates a form of straw feminist. People with an interest in revising the associated gender scripts, including the ways iPhones cannot be used by men, could work with such feminist critiques and encourage further inclusion rather than repressing the debate.

There is a cultural tendency to suggest that individuals will (and should) find ways of accommodating themselves to the gender scripts of devices over time. The repressive reporting on Watson-Currie's and other people's critiques, as I argue in this chapter, is designed to encourage women to stop insisting that their fingernails should be acknowledged and should be included in device design. More generally, the dismissive reporting and commenting are meant to patrol gender norms. A quick review of technology forums might suggest that such didactic practices, including dismissive indications that women should cut their fingernails, result in women refashioning their self-presentations, downplaying professional expectations about "nice nails," and finding that iPhones are no longer a problem with shorter nails. However, an analysis of more recent blogs, forums, and women's magazines suggest that women continue to identify problems with capacitive phones that are intensified because of their fingernails. For example, Angela Sitilides shares her thoughts on trying the "long nails" trend in 2019 and juxtaposes fashion and device configurations. She chronicles that texting "has been a challenge" and searching "isn't quite as quick" as her previous experiences.[102] Carly Cardellino's "15 Things You Super-Annoyingly Can't Do With Long Nails" from 2016 describes how "you have to change your typing ways and use the pads of your finger more" and "be prepared to hit the wrong keys ALL THE TIME. alsk@%$@#!"[103] Her examples convey how typing with fingernails results in and meshes errors and frustration. She emphasizes the significance of these experiences by subtitling the article, "The struggle is real, y'all." However, Cardellino does not consider how device designs might address rather than amplify these occurrences and the associated sentiments.

The Stylus for Fingernails

Forum commenters often respond to women's critiques of the design of iPhones by indicating that women should rethink their embodied positions. For instance, people "suggested typing on the side" of her fingers to Noelle Buscher, but it is "hard to be accurate."[104] Ken Levine proposes in his blog that women "could touch the screen with the side of your finger, or the pad of your finger, or your thumb, or new nose?"[105] Levine dismisses the addressed women as too aesthetically involved (and having employed cosmetic surgery), while using photographs that objectify women. These unconventional and sometimes facetious recommendations displace cultural figurations of hands and fingers as dexterous. Such suggestions, which are unlikely to result in speedy and facile engagements, often combine the gender scripts of mobile devices with a more general dismissal of women's bodies and experiences. Yet individuals' proposals for women to adjust their screen-based typing methods do not account for the ways nails and fingers hide a portion of the screen. As Lindawcca indicates, "it's impossible to even see the 'pop up confirmation of the letter/number you selected."[106] Women with fingernails are scripted as ignorant, but the people who chastise them do not understand.

A variety of companies offer peripherals that are meant to address such quandaries and provide additional ways of engaging with touchscreens. For instance, I have previously noted how Apple promotes its pencil as an ideal tool. Other people address the sometimes difficult connection between devices, styluses, and their hands. The SolidSmack site indicates that individuals who are challenged by capacitive screens span from the "the calloused-handed carpenter and the sodden-fingered dishwasher, to persons struggling with arthritis or Parkinson's disease."[107] SolidSmack concludes that "we have all witnessed the struggles associated with touching small screens, and most of us have had our fair share of infuriating moments." The site emphasizes bodily differences and underscores how device glitches produce feelings. Rather than offering accommodations for all bodies, digital device designs persist in highlighting and even producing bodily disability, which includes carpal tunnel syndrome and other sorts of corporeal pain. SolidSmack adds, "Aside from gloves, it is probably safe to bet that long fingernails are the most common nemesis of the touchscreen, and for that reason, Dermatologist Sri Vellanki personally funded her brainchild innovation to give the world what mother nature could not: stylish stylus fingernails!" Thus, SolidSmack and Vellanki propose a counternarrative to Jobs's insistence that fingers are the natural and best pointers for everyone.

SolidSmack and Vellanki could fully advance an alternative gender script by associating women and fingernails with thoughtfulness and innovation. However, Vellanki and many of the journalists reporting on these peripherals risk depicting nails as a kind of glitch in the bodily system that can be resolved with technological intervention. Vellanki offers a nail-like stylus, which slips over the fingertip, and a prosthetic nail and nail tip that are attached in the same manner as press-on nails. In the associated illustrations, a woman employs her burgundy nail to accomplish varied screen tasks. The development of nail styluses promises to modify women's experiences with touchscreens while, in a manner that is related to reporting on women and capacitive screens, still rendering women and their fingernails as problems. Amanda Kooser reviews Vellanki's Nano Nails and notes, "Long nails and touch screens don't get along very well."[108] As part of Vellanki's description of the stylus, she identifies "problems" with "hitting the target" because of her nails.[109] Vellanki "had to work around" her "nail to hit the screen." Her evocation of working around the nail articulates difficult bodily relationships and distinguishes between the nail and the targeting finger. It is as if Vellanki could move, or step, around her nail. Many Internet commenters suggest a similar moving around and altering of the body when they argue that women who have problems with capacitive touchscreens should cut their fingernails or stop using the technologies, rather than considering that technologies can be adjusted to facilitate people's needs.

Vellanki associates the fingernail stylus with women. Yet teepingpom notes that the hand in the video, which demonstrates the peripheral, "looks like a man's hand."[110] "Disgusting," writes Chop-Sue-Me, "Do you know how dirty long nails are? It also makes you look like a dirty crack whore. Not sexy at all."[111] These individuals resist the technologization of women's hands and suggest that the product and associated women fail to meet men's desires, even though the product is designed to facilitate dexterity and activity rather than eroticization. These comments are troubling since the individual demonstrating the technology is a woman of color. The commenters stereotype her and more generally represent women of color as not feminine, clean, or respectable enough. These posters thus deploy the kinds of incivility mentioned by Sobieraj and Berry and correlate women of color with addiction and sex work as methods of degrading them.[112] Terms like "disgusting," in a similar manner to the earlier texts that I analyze, imply that the depicted hand and associated woman are not fully human. The commenters reference normative hands as a means of distinguishing and elevating white people. Their narratives thus work with many digital representations of hands, including the hand-pointer.

Chop-Sue-Me and teepingpom correlate women of color with the monster, even though Vellanki's Nano Nails are designed to resolve embodied differences and the ways capacitive screens create problems for people with longer nails. These renderings of women of color as inhuman is a common racist convention that I mention in the introduction and in the previous section. This association is then intensified by commenters' references to bodily horrors, including feelings of disgust. For example, the writer Buster Hein references horror texts and unstable and mutable bodies in "Your Long Fingernails Can Now Be Transformed Into Touchscreen Styluses."[113] While he has "no idea what it's like to try to use an iPhone when you have super long fingernails," he still feels qualified to describe women trying to "stab away at their screens like they're Freddy Krueger." Hein thus articulates the status of women while distancing himself from them and feminine behaviors. Dan Moren similarly reports, "We're not just talking Freddy Krueger or Sabretooth here, either. Some women with longer fingernails are accusing Apple of misogyny."[114] Moren thus hints that women's critiques of intolerance are more unacceptable than the murderous Krueger character in the 1984 horror movie *A Nightmare on Elm Street* (directed by Wes Craven) and the ongoing sequels.

Hein and Moren render women as pretechnological monsters that cannot employ touchscreens. Given the prevalence of narratives about bodily horror in everyday culture, as I suggest in previous research on wedding cultures, an attention to such narratives and an employment of the critical literature on this topic is productive.[115] For instance, Carol Clover has noted that monsters, especially murderers in slasher films, ordinarily kill with their embodied features and extensions rather than technological devices.[116] As Clover and other horror scholars suggest, the monster is removed from binary gender and functional heterosexuality, which correlates with reporters' and commenters' scripting of women with fingernails as unrecognizable and not satisfying men's sexual interests.

Hein and Moren collapse the monstrous male and phallic body of Krueger with women's fingernails. Hein proposes a purportedly better bodily configuration where women can use Nano Nails to "stop looking so weird when you're trying to send a text message." The Nano Nails device is imagined to better align women's bodies, to make them presentable, and to straighten them out. Like journalists' reporting on iPhones, these technology writers and commenters reference Nano Nails as a means of reshaping women's hands into a form that is more acceptable to the appraisers. Thus, the comments about the Nano Nails attempt to constrain women's hands in the manner that Ahmed identifies. Commenters try to refuse women's interventions into iPhones and retention of the specificity of their hands. In the next section and chapter 4, I continue to address the ways women recraft normative cultural demands about their hands and fingernails so that these physiognomic features favor their own aesthetic and political interests.

Conclusion: The "extra oomph" of Fingernail Extensions

Society often distinguishes between men, who are imagined to have bodies, and women, who are thought to be bodies, to be available for the male gaze, and to be too focused on their own appearance. In previous research, I demonstrate how the scientist Hope Jahren and her #ManicureMonday hashtag hijacking have addressed gender norms, but also worked against women's beauty cultures.[117] Jahren encouraged scientists to post images of their fingernails to Twitter and to use the #ManicureMonday hashtag as a means of countering what she deemed to be frivolous nail polish applications. She encouraged women to self-identify with women scientists' working bodies and to thereby refuse beauty aesthetics, even though the working body is an aesthetic. Nano Nails provide some intervention into these distinctions when it advertises "technology and beauty combined for the modern woman."[118] Yet Emily Blake adopts a similar position to Jahren when she responds to critiques of iPhones by writing, "Any woman who wears her nails that long KNOWS it makes simple tasks more difficult. That's the choice you make to have a fancy manicure."[119] Blake employs a version of "it's her fault" and suggests that women have made bad choices that take away their rights to interrogate issues. She further establishes gender scripts by arguing that her "hands are made to do stuff, not look pretty." Jahren has asserted, "it's not about how your hands LOOK it's about WHAT THEY CAN DO."[120] These individuals' deceptive assertions about choice and the best ways of self-presenting, which try to dissuade women from choosing to have longer nails and nail applications, fail to acknowledge different types of embodiment. Individuals' beauty practices are a form of doing stuff that is influenced by normalizing cultural pressures and individuals' desires. To choose some identities or objects should not prevent individuals from interrogating the ways they are held back from inclusion in cultural and design decisions.

Women with fingernails are often dismissed for their aesthetic interests, as I demonstrate in this chapter, which pushes their hands and nails away. Academics have also ignored this online debate and therefore elided some of the gender scripts and normalizing features of iPhones and related technologies. In overlooking such online contestations, scholars risk supporting male-centered views of technologies that control and dismiss feminine aesthetics. However, reporters and commenters, as I indicate in chapter 2, often celebrate iPhones because of their aesthetic features. In that chapter, I consider how the touchscreen is figured as an aesthetic device and as a sometimes troubling skin. For instance, Allan Sampson contrasts women's fingernails with human touch and the human body when writing, the "touch screen requires skin."[121] In a discussion about "Women's Nails vs. iPhone," eastercat advises, "It's a *touch* screen. It reacts to skin."[122] These commenters indicate that the touchscreen figures and requires a specific kind of body, which is not facilitated by fingernails. Of course, as some commenters have replied, nails and other body parts are capable of and are often employed in touching.

Women's accounts of using resistive screens and related devices with their fingernails also contradict these descriptions of touchscreens as skin oriented. For instance, adebaybee identifies a "Samsun Omnia" as "99% responsive to nails poking."[123] She previously used a Hewlett-Packard (HP) device with a screen-based keyboard and "was able to poke w/o having to use the stylus." ljahnke has a "cheap LG Chocolate" and "can type on the keyboard" with "fake nails."[124] She asks, "Couldn't Apple offer a choice of screens for those of us that have fake nails?" Women speak back, raise their arms, and assert the functional aspects of fingernails in narrating the ways their bodies work with other devices. They assert that earlier and less expensive devices were more accommodating and useful. These women thus propose the utility of embodied extensions and fingernails. Like Clover's analysis of monsters and "final girls," who fight with embodied extensions, these women with nails render a nonnormative and active feminine body that works with and through technologies.[125]

Women's arms are unfortunately pushed down when Julian Wright employs an exaggerated version of their narratives to mock them. He "can't believe that Apple can't offer a screen that can be used with fake beards."[126] He has a "cheap LG Chocolate" and "can type on the keyboard" with his "fake beard. Couldn't Apple offer a choice of screens for those of us that have fake beards?" Wright repeats and revises phrases as a means of undermining women's critiques. He thereby engages in a version of what Barbara Johnson describes as "incompatibilities between explicitly foregrounded assertions and illustrative examples."[127] By referencing and rethinking Johnson's reading methods, I want to suggest that Wright engages in a persistent form of straw figuration. He and other individuals pretend to repeat and provide examples of texts while actively revising them as ways of undermining the associated arguments. Wright employs such an approach to focus readers on what he sees as the normative design and version of men's bodies, and through this process, he asserts the rights of men. Women provide a critical and political problem that he works to diffuse when their narratives about "poking" disrupt the gender and genital order enforced by iPhones and the people who use them. Thus, feminist hands work to reconfigure hierarchies and bodily organizations.

Some of the comments considered earlier in this chapter, including Watson-Currie's and adebaybee's texts, dispute indications that fingernails are incapacitating. In Annie Kreighbaum's article about growing her fingernails, she argues, "Long nails not only visually lengthen your fingers," but also provide that "tiny bit of extra oomph you need to grab your phone from the other side of the table, or reach for the other handle across a fully-stuffed mesh laundry hamper."[128] The hand with fingernails may thereby be better at grasping and employing some objects. Kreighbaum extends this argument when indicating that nails are physical prosthetics that allow people to perform otherwise undoable things. For Kreighbaum, the "oomph" of long fingernails includes visual augmentation and extended reach. It can also result in chips and painful tears. Devices and hardware are coded as extensions and aspects of people's bodies through buttons, bumps, and impressions. Some mobile phones include fingernail ridges that indicate that individuals can open the phone and thereby render fingernails as always ready-to-hand tools.

Instructions on how to open up earlier phones often refer to fingernail slots and grooves. For instance, Tanker Bob reviews the Samsung t629 and advises that the "slider opens easiest by hooking your thumbnail in the slot above the screen and pulling with the other hand on the outside rim of the phone."[129] Wilson Wong's review notes that the Nokia requires similar embodied configurations, and "unless you have long fingernails or even a spoon, trying to dig your thumb in to open the clamshell phone is difficult, if not impossible."[130] As these reviewers begin to suggest, fingernails are required tools for some phones and actions. These designs and texts indicate the ways nails extend bodily functions and underscore the prosthetic aspects of embodiment. Technological extensions are often embraced features of contemporary society, including individuals' online representations, the shifted conceptual reach of people's Internet-facilitated communications, and the attachments that people have to digital media. Academic and popular readings of Donna Haraway's cyborg manifesto identify technologies as prosthetically connected to and extending individuals' bodies.[131] Mobile devices and cars are frequently incorporated into people's understandings of the self and may be understood as bodily extremities.

The literature on gender scripts provides useful ways of engaging with technology designs and intervening in the identification of digital and other technologies as unbiased tools and bodies. The iPhone's gender scripts distinguish between men's and women's bodies, technological capacities, intelligence, rationality, and humanness. Normative men's empowered positions are buttressed by the design of the device and people's refusal of women's critiques. Commenters further maintain this traditional societal structure through name-calling and other containment strategies. Sobieraj and Berry's research suggests how such methods coerce women and enforce the status quo. These tactics, as I note throughout this book, also render straw versions of feminists and women. Critical texts and popular intercessions, including the practices that I outline and develop in this book, are thereby important methods for studying digital culture and intervening in the experiences of disenfranchised subjects.

People's identification of iPhones as intuitively usable and women's critiques as dated and trivial are methods of denying that men's bodies and expectations are empowered by designers, manufacturers, and devices. They also conceal the larger ways these texts are used to perpetuate traditional identities and hierarchies. Due to such controlling practices, academic and popular analyses of gender scripts should include interrogations of claims about addressing "everyone." As Rommes suggests, gender scripts mandate culturally denigrated subjects to change because these individuals do not (and by design never will) meet cultural norms.[132] While the literature underscores how designers and companies produce notions of users, my study indicates the everyday ways journalists and commenters render technologies. This includes their association with and support of brand scripts. My study also points to the need for academics to further consider the ways individuals and groups support gender scripts. Without such studies, key facets of gender scripting and their cultural influences will continue to be normalized and elided.

Ahmed's analysis of the raised arm provides methods of thinking about what happens when individuals question gender scripts.[133] In her figuration, the raised arm will not consent to cultural mandates and will not change its insistence on being recognized and maintaining its own position. Raised arms and hands can function as protests and can refuse the schemas that digital devices map for bodies. They do not offer their appendages for use by managers, companies, and the state. I extend Ahmed's notion of the raised arm and employ it and the raised hand and fingernails as theoretical devices when thinking about computer and online systems, because the hand is presumed and represented in these settings. Scholarly considerations of the limits of raised arms are also needed because, as this chapter demonstrates, individuals rescripted Watson-Currie and other women's raised arms as means of maintaining cultural and technological hierarchies. Hand-pointers, and thus key features of operating systems and interfaces, keep hands in line. The designers of these representations and methods of engaging presume that individuals are able to present their hands in certain configurations and physically and conceptually take up mapped positions. In a related manner, disability scholarship has shown how deaf and hard-of-hearing individuals are expected to enact voice phone call norms even when they are employing asynchronous accessibility options.[134] The designers of hand-pointers and iPhones articulate a normative subject who uses technologies. However, they also depend on women with fingernails, people of color, and disabled subjects as methods of indicating that these cohorts are not ideal (or not participants) and advancing scripts that work against these constituents.

2 The "interface, represented as a skin": Oleophobic Coatings, Touchscreen "Scars," and "Naked" Devices

Women with longer fingernails, as I indicate in chapter 1, are often depicted as frivolous and as having chosen their aesthetic appearance over technological proficiency. In such cases, fingernails and the associated women may also be understood as dirty and contaminating. While fingernails function as screens, as I suggest throughout this book, they are also figured as hindering people's employment of capacitive screen technologies. Fingernails screen people's nail art, function as ready-to-hand tools, figure the position of individuals and their hands, and cover the soft tissue under their surfaces. Fingernails can also screen technologies from the embodied traces that people leave on touchscreens. However, this requires screen technologies that are resistive (and respond to pressure) rather than capacitive (and reliant on skin's ability to conduct electricity). These and other narratives about cleanliness and dirt, as I suggest in this chapter, are part of people's ongoing conceptions of mobile, touchscreen-based devices. While some individuals identify dirt and damage as records of their positive engagement, many people lament and mock dirty and damaged screens. They also relate varied device films and skins to embodiment, thus further meshing technologies and individuals. Filth and failure are also too often correlated with women and femininity. I critique the ways people on technology forums focus on the aesthetics of screen technologies and touching but dismiss women for being interested in nail aesthetics and other beauty cultures.

Screens are archives of individuals' touch. For instance, iphonefreak450 describes having "fingerprint and smudges" on the iPhone "screen since it is so humid."[1] Despite "cleaning it with a microfiber cloth," iphonefreak450 still gets "this film all over" the screen. So iphonefreak450 asks for assistance in cleaning and maintaining this screen. In a similar manner, AmazingTechGeek wants to know how to "avoid fingerprints/smudges from attracting on an iPhone."[2] These forum posters see their bodies and oily residues on screens and chronicle them for a readership with similar experiences. At the same time, as they express some level of concern about embodied experiences with screens, their member names assert technological proficiency and relationships with mobile devices. This may provide them with some separation from the described screens, which are coated with a "film" and provide imprints and versions of the associated individuals' skin. While touchscreens offer versions of the individual, which are related to how computer technologies are understood as a form of the owner and as addressing the ideal user, these dirty and damaged screening surfaces and the associated residues are ordinarily identified as undesirable, tainting, and disgusting. They are correlated with the stickiness of humid weather and the damp bodies that imprint themselves on shiny and smooth surfaces. Screens can be records of other everyday but disagreeable bodily practices. Yet dirt is distinguished from and purportedly repelled by the oil-resisting oleophobic screen coating that I consider later in this chapter.

Applejuiced correlates screen smears with the supposedly uninformed act of eating a "cheeseburger or a slice of pepperoni pizza" and then wondering "why there's smudges" on the "screen and other peoples iphone's are cleaner."[3] Newtons Apple also takes a dismissive position when identifying "people eating and using their phone at same time. No way those French fries are not smeared on the screen. Just wash" your "hands several times a day."[4] The associated screens are supposed to show the unhealthy and unaware actions of such individuals. For instance, Newtons Apple dismisses purportedly uncontrollable and unreasonable fat by rendering the initial poster as a child who is in need of control and instructions about handwashing. Individuals are imagined to be producing and desiring greasy devices (and foods), which continues stereotypes about uncontainable bodies and excessive consumption. In her fat studies book *Fat Shame: Stigma and the Fat Body in American Culture*, Amy Erdman Farrell indicates the "enduring power of fat stigma; the way fat denigration overlaps with racial, ethnic, and national discrimination; the connections between both of these (fat and ethnic denigration) and class privilege and, finally, the ways that all of these elements" contribute to a "properly gendered subject."[5] People's figurations of unacceptably fat bodies and greasy devices thus enable them to articulate embodied norms and clean devices that recognize their touch and expertise.

Applejuiced, Newtons Apple, and related commenters establish their more culturally acceptable embodied practices. They also convey their potentially excessive Apple fandom through member names. Given the multiple things associated with apples, they risk rendering themselves as food products as a means of self-identifying as scientific apples, including references to Isaac Newton's experiments and the fact that Apple named one of its personal digital assistants the Newton. These commenters' narratives about messy and disordered eating in front of and on screens are related to the dismissive figurations of inactive, gluttonous, and fat computer users that I critique in *The Body and the Screen*.[6] People understand fat as contaminating, and as differently coding men and women, because of its association with femininity. These chronicles about dirt are also associated with ongoing fears of viral contamination and COVID-19, which I analyze in the afterword to this book.

In this chapter as well as the entire book, I extend my considerations from *The Body and the Screen* and study the ways normative male bodies, including not-too-fleshy or -greasy bodies, are elevated and privileged through screen engagements. My earlier research notes the ways screens are effaced in order to make it seem as if empowered male bodies can move because of the affordances of digital media and operate as part of online representations. In this chapter, I focus on the specific ways people correlate devices with versions of individuals and corporeal skin. Mobile touchscreen-based devices are further enlivened when they are wrapped in and thought of as skin, including human skin. People associate these broad notions of digital and human skins and connections with sensations. Individuals articulate pleasurable feelings about and attachments to undamaged devices and uncomfortable engagements with screen technologies (and thus the self and the other) when screens are cracked and thought to be otherwise contaminated. In a similar manner to the ways that theories of spectatorship indicate how the ideal viewer is produced as a means of regulating nonnormative individuals, I consider how the production of the skin and body of the mobile phone and its owner render normative users and specify how individuals should appear and act. This includes articulating normative men as intended users and employing women as methods of explaining devices and objectifying and dismissing women's bodies.

My inquiry in this chapter is informed by scholarly considerations of the body, which have been prevalent in the last few decades. This includes body studies scholars' generation of and association with skin studies. I employ the scholarship on skin, including theorizations by Sara Ahmed, Didier Anzieu, Nicolette Bragg, and Naomi Segal, to analyze how devices become conceptually intermeshed with and impressed in individuals' flesh.[7] Body studies researchers address such things as the ways the body is correlated with gender and other identity positions and less valued than the mind and thinking. They foreground sensations in considering how the body is experienced through a variety of feelings and is deemed to be disgusting and too fluid, especially when it is associated with women and other denigrated subjects. Bodies are thought of as more ideal when they appear to be coherent and impermeable. The properties of skin, as Steven Connor indicates in his book on the topic, have often been ignored as a means of conceptualizing the body as a delimited object that concludes with and is defined by the skin.[8]

Connor does not connect methods of foregrounding skin, and thereby ways of advancing more fluid and expansive notions of the body, with the digital. However, skin is correlated with digital technologies and devices through the common identification of interface options and mobile phone cases as "skins." In this chapter, I consider how the cultural articulation of touchscreen-based phones and other devices as skins intensifies the ways people associate these devices with individuals' bodies and feelings. I analyze the similar online formulations of touchscreens, especially notions of oleophobic coatings, dirty devices, and cracked screens, that appear in blogs, manufacturers' websites, news articles, patent literature, and technology forums. These texts conflate dirty and cracked phones with skin and script the associated individuals in relation to gender norms. While women's phone use tends to be culturally denigrated, especially for women with longer fingernails, men's aesthetic interests in skinlike phone cases and other peripherals are encouraged in technology forums and the other sites that I study. People's correlation of culturally acceptable skin, bodily feelings, and digital technologies, as I argue, results in people managing dirty and damaged devices. While some people render themselves as composed and rational bodies by indicating their lack of concern about the condition of mobile devices, and positively self-assess in comparison to more emotional users, individuals tend to be concerned about how technologies challenge their identity.

Skin Studies

Skin studies, as the sociologist Marc Lafrance indicates, focuses on the surface of the body and is informed by some of the same critical interests as body studies, including concerns about identity, power, and cultural classification.[9] Skin studies also troubles the notion of skin as a container and as a singular, thin wrapper. It considers the ways the surface of the body is tenable and understandable. Research in this area also understands skin to be "processual, relational and sentient"; "human and non-human, material and immaterial, indeterminate and multiple"; and "bound up with thinking and, indeed, rethinking agency, experience, power, and technology."[10] Lafrance notes that skin is a persistent aspect of people's experiences and always changing. He underscores the importance of skin as individuals' largest sense organ (and the largest organ more generally), the heaviest part of the body, and a key component of sensing and living. He also mentions the French psychoanalyst Didier Anzieu's similar arguments, which I employ in this chapter.

The term "skin" is commonly used to convey a variety of states through such phrases as "skin and bones," "soaked to the skin," "skin in the game," and "skin of one's teeth." People understand the body and the world through these phrases and through skin. Lafrance describes skin as the "frontier of inside and outside" and "self and other, subject and object."[11] Skin is thus in between. This in-betweeness is related to Lafrance's description of skin as a "fluid boundary and a leaky interface."[12] Skin is not seamless, being constituted of wrinkles and openings that enfold and drive things into and out of it. As I suggest later in this chapter, Julia Kristeva's theory of abjection references different types of film and skin as examples of disgusting things and as properties that must be expelled to retain a notion of self.[13] While skin can be experienced as an in-betweenness and a series of opposites, including such binaries as delicate and resilient that are part of mobile phone owners' narratives, it can also be perceived as a border, a shield (or screen), and protective armor. Skin is believed to bind the individual into a comprehensible whole and to distinguish the individual from other people and things. The cultural insistence on distinguishing between the self and the other must manage conceptions of the body and skin as open and fluid. Lafrance's description of skin as a leaking interface references the imperfect boundaries that can never fully be controlled and the ways skin and fingers are conceptualized as a conduit, including a connection to hand-pointers and a form of digital interface.

Anzieu uses the term "Skin-ego" to convey the ways people are defined by and experience other people and the world through their skin and tactile relationships. Anzieu highlights the vital role of skin in identity construction. According to his psychological account, the child comes to understand itself and the world through the processes of mothering and the associated physical engagements with people. Anzieu's concept of Skin-ego refers to a "mental image used by the child's Ego during its early stages of development" that contains "psychical contents, based on its experience of the surface of the body."[14] In Anzieu's account, the skin is experienced as multilayered and enacted through varied kinds of cutaneous contact. The mothering atmosphere encases the baby and includes an "external wrapper made up of messages."[15] The "double feedback" between the inner and outer wrappings and between other people, especially the mother and child's engagement with and production of skin, results in an "interface, represented as a skin common to the mother and the child." They are "'plugged in' to each other through the common skin" and thereby "communicate directly, with reciprocal empathy and an adhesive identification: it is a single screen that resonates with the sensations, affects, mental images, and vital rhythms of both." Anzieu's description of a shared skin provides a method for understanding the connection between individuals and devices, including the ways mobile phones are incorporated into people's skin and identity and the shared surfaces and experiential interfaces that result.

Anzieu refutes the notion of the skin as a border that protects and leads to an essence or central core. Naomi Segal, who has translated and analyzed Anzieu's work, suggests that the "skin has a double surface."[16] Anzieu describes a protective skin on the "outside and, underneath it or in its orifices, another layer which collects information and filters exchanges."[17] Segal indicates that this conception of complicated and communicating surfaces can assist people in comprehending the corporeal, psychical, and cerebral realms in a different way and replace the more traditional explanation of how thinking penetrates an essential self and establishes notions of truth. Such conceptions of surfaces and layers offer methods of theorizing the relationship between individuals and devices and their shared and impressed skins. I employ Anzieu's and Segal's research and recent feminist scholarship to underscore the ways skin is collaboratively produced and influenced by others.

The feminist research of Nicolette Bragg theorizes this enmeshed and resonant skin through her own bond with her infant child. She notes that her "daughter's body is never without another, supplementary surface" and that bodies in contact produce indents.[18] Bragg proposes the concept of "beside oneself" as a means of understanding the mother's position in child development and the ways people experience their bodies relationally. This concept "describes the non-possessed self marked by the possible and creative re-formulation of one's own contours into another's boundaries."[19] Employing this and related theories, as I do later in this chapter, the individual using the mobile phone can be conceptualized as experiencing the body beside itself as the person is the phone *and* the user of the phone. The phone pushes, vibrates against, and claims to be a version of the body as the individual is activated through these signals, accepts the associated calls and embodied configurations, and resists these states.

Ahmed's indication of how feelings influence understandings of inside and outside is related to Bragg's analysis. According to Ahmed, "to say that feelings are crucial to the forming of surfaces and borders, is also to suggest that what makes those borders also unmakes them."[20] She suggests that instead of thinking of the skin surface as a container, that skin should be understood as the site "where others *impress* upon us," which might include the ways mobile phone screens can become impressed with the texture of people's clothing, stuck in pockets where they indent skin, and scratched by belongings.[21] This notion of impression is similar to Bragg's narration of bodies indenting other bodies. It evokes the emotional, willing, and coercive features of skin contact and molding. Ahmed encourages individuals to "unlearn the assumption that the skin is simply already there, but begin to think of the skin as a surface that is felt only in the event of being 'impressed upon' in the encounters we have with others."[22] This includes the ways individuals read and form the skin of different people. Thus, communities and things are participants in and co-constitute physical and emotional feelings.

Apple produces a version of these intertwined sensations and skins when noting that the "textured back glass" of its 11 and 12 generation iPhones "provides an elegant look that is also tough, slip resistant, and feels good in your hand."[23] With this text, Apple continues to emphasize the connections between design, aesthetics, and embodied fit and pleasure, which, as I suggest in chapter 1, are not delivered or addressed in instances where women cannot use iPhones because of their fingernails. Women's bodies and methods of skin contact are expected to change for the phone, while men's associations are allowed to be more malleable and modify the device. As Apple has indicated, the mobile device's "textured glass may show signs of material transfer from objects that come in contact with your iPhone, such as denim or items in your pocket. Material transfer may resemble a scratch, but can be removed in most cases." Thus, iPhones act as a version of Lafrance's leaky interface where hands, other body parts, clothes, and the world slip in and impress themselves on the phone. As Bragg similarly suggests, it is a relationship where the individual is beside oneself and produces the object through a series of indents and transferences that are the body and the phone and that are removed from the body and the device. People's experiences with impressions combine close readings of device surfaces with critical proposals for how such arrangements function. I thus employ these theorists' considerations of skin as a means of considering the relation between the body and device and how online posters' descriptions of devices influence other participants.

A growing number of contemporary film and media theorists account for the skin and body in their studies of production and viewing. In Vivian Sobchack's phenomenological scholarship, which I mention in the introduction, she describes watching the opening of the 1993 movie *The Piano* (directed by Jane Campion) and feeling the scene with her body and fingers. Since the scene is blurry, she cannot immediately identify a point-of-view shot that depicts the world through the fingers of the character. Yet her "fingers *comprehended* that image, *grasped* it with a nearly imperceptible tingle of attention and anticipation."[24] Sobchack indicates how something is cognitively "grasped" by being felt in the individual's fingers. The term "grasping," as Sobchack intimates, combines tactilely touching and knowing. She thereby chronicles how her skin responds to and is influenced by the film. The ways film texts render forms of skin contact and reshape bodies is expanded with touchscreens, which are often so closely in contact with individuals that they indent bodies and amplify emotional connections.

Laura U. Marks's *The Skin of the Film* similarly suggests how "vision itself can be tactile, as though one were touching a film with one's eyes."[25] Touchscreens further intermesh conceptions of seeing and touching, and sometimes displace seeing with feeling. They render touch as a form of seeing and knowing. Touchscreens also make fingertip contact, as my introductory analysis of N. Katherine Hayles's comment about digital research indicates, into a form of reading.[26] Marks argues that conceptualizing film as skin "acknowledges the effect of a work's circulation among different audiences, all of which mark it with their presence."[27] This marking is personalized with digital devices, as I elaborate in the following sections, and becomes records of individuals' engagements. This includes people noting pleasurable marks of use in the form of stickers and familiar rough spots. Individuals also experience their cracked screens as painful scars that continue to abrade skin when they rub against these surfaces.

People imagine unmarred phone skins and surfaces and identify "good" corporeal skin as smooth and pale (but not wan). However, theories of skin assert the varied ways skin is experienced, its myriad surfaces, and its plethora of functions. As Steve Pile notes in his research on skin, "More than a container or boundary layer or frontier even, the skin ego is lumpy, misshapen and unevenly developed."[28] Lafrance evokes a seamed skin and related notion of self that has pores, orifices, and other cavities that are open to and enwrap people and the world.[29] Seams and scars are sites of rupture and contention that may be worriedly picked at. People may more neutrally trace the seam between parts of cellphones or other handheld digital devices. The skin that is shared by the person and touchscreen is sometimes also tentative and torn by attractions to and rejections of other bodies and surfaces.

People associate broken skin with abjection and related feelings of bodily disgust. Kristeva identifies abjection as things and experiences that disrupt identities and systems and that do not respect boundaries, situations, beliefs, and rules. According to her, food revulsion is the most basic sort of abjection. In her account, such unexpected and improper skins as the film over milk cause individuals' distressing encounters with the other and experiences of disgust. When Kristeva sees or her "lips touch that skin on the surface of milk—harmless, thin as a sheet of cigarette paper, pitiful as a nail paring," she experiences a "gagging sensation and still farther down, spasms in the stomach, the belly; and all the organs shrivel up the body."[30] Kristeva identifies the ways thin films repulse individuals. They barely cover surfaces but act as a kind of caul when fluids are engaged. The associated ruptures in films and skins challenge notions of unitary and consolidated subjects. For instance, Kristeva asserts that leprosy "visibly affects the skin, the essential if not initial boundary of biological and psychic individuation."[31] She also risks articulating a coherent and consolidated body and rejecting differences through this account. The coronavirus and other communicable illnesses further disrupt notions of individuation and render and fracture communities. The coronavirus encourages and interrupts people's desires to feel and touch, as I suggest in the afterword, by entering, exiting, and connecting bodies. The spread of viral particles and fluid droplets sometimes results in COVID-19 and the catastrophic death of individuals and groups.

Fluids, according to Segal's study of Anzieu and skin, are disrupting agents. Body fluids "breach the bounds of the body" and "provide the skin-ego's grievous inability to contain, they become both dangerous and ambiguous."[32] The feminist philosopher Elizabeth Grosz chronicles attempts to shore up this open body and argues that the "obedient, law-abiding, social body" is developed through directives about body fluids.[33] There are also concerns, especially when referencing men's corporeality, about anything that breaches, penetrates, opens, and liquefies the skin and body.[34] It would also be productive to consider dry, grained, and cracked skin that separates at points and hangs loosely over internal supports and viscera. In this chapter, I consider such embodied, fluid, and digital surfaces and indicate how skin studies and related research elucidates gendered conceptions of mobile touchscreens. I consider the ways people identify touchscreens, including oleophobic coatings and dirty and broken surfaces, as sites of feeling and personal skins. Individuals who use mobile, touchscreen-based devices are linked to their hardware through expansive notions of skin and often repulsed by the filthy and broken skinlike features of touchscreens.

Gender, Dirty Hands, and Oleophobic Coatings

People frequently conceptualize Apple's oleophobic coating as a protective and delicate skin, which they worry about damaging. Apple introduced and began marketing the oleophobic coating with the 3GS.[35] According to ajinkya@tmrresearch.com, the term "oleophobic" "refers to materials with zero affinity to oil" and the coatings are "ideal for preventing fingerprints on smart device displays."[36] People's explanations of oleophobic coatings are often contrary accounts of smudge-free displays and cleaning the screen.

They are also articulations of the connections between individuals' dirty hands, skins, and devices. Thus, PaulK offers an article titled, "Oleophobic coating – what it is, how to clean your phone, what to do if the coating wears off."[37] He also relates and distinguishes bodily traces from screen surfaces when noting that the coating is a form of "finger smudge resistance." This oleophobic coating is represented as skin or, as PaulK indicates, a "layer." He recommends against cleaners, which can "easily wipe off the oleophobic coat and leave your glass 'naked.'" This notion of the "naked" links digital devices to corporeal embodiment. The tactile features of the caseless, "naked" phone are celebrated, and its fragility is worried over, as I indicate later in this chapter. These frameworks, in a similar manner to the cultural conceptions of oleophobic coating that I consider in this section, employ women's bodies and stereotypes as a way of explaining and coding touchscreen devices.

In a post that is supposed to explain the oleophobic technology, the science writer Bill Nye uses an analogy about water droplets sticking to the "nylon fibers in a bikini strap, the swimsuit feels wet (or so I'm told). When they don't stick to the surface they're resting on, they bead up, like in the car wax commercials."[38] Women's bodies and clothing are thus employed to explain the features of the oleophobic coating, but it is other things, such as waxed cars, that Nye associates with desirable properties. It is worth noting that "waxing the car" is a metaphor for male masturbation.[39] Since the bikini is also culturally figured as one of the clothing items in which women erotically appear, these swimsuits help produce what Laura Mulvey identifies as women's proscribed position as to-be-looked-at-ness.[40] This composition of women as the object of the look is combined with the constitution of them as to-be-touched-ness, as I have outlined in earlier parts of this book. Women are rendered as to-be-touched-ness because of the materiality and tactility of Nye's formulation.

Nye articulates women as touchable while trying to distance himself from feminine experiences. He asserts his intellectual rather than prurient interests by denying that he knows how a wet bikini strap "feels." This positioning of women is underscored in reports about this post. For instance, Bryan Chaffin identifies Nye's explanation as "fun and easy to understand" and renders women's bodies as pleasurable and simple cyphers.[41] The commenter ch3burashk highlights the ways Nye frames women as objects when writing, "Nylon bikinis? Wet (or so he's told)? Bill is a freakin' player."[42]

ch3burashk foregrounds the ways Nye buttresses his masculine position and uses the narrative as a means of establishing a shared site of male heterosexuality and technological homosociality. Technology designers, marketers, manufacturers, and consumers produce gender scripts, as I suggest in more detail in chapter 1, when they identify expected participants as men and employ women's bodies as mere frameworks to explain the ways individuals should understand and engage with devices.

Other commenters on Nye's article assert their masculine position. I closely read such texts throughout this book as a means of demonstrating the ways participants collaboratively produce their identities and beliefs. For instance, wiggin produces a gender script when equating the iPhone's scratch-resistant glass to men's association with ruggedness and unfettered sexuality. wiggin argues that safeguarding the phone and oleophobic surface with a "screen protector was like getting a vasectomy, a full testicular removal and then ten years later putting on a condom just to not knock her up. Totally unnecessary."[43] This commenter understands the more durable iPhone as a corollary for the male body, and specifically for men's sexually active genitals and procreativity. Any restraining of genital and sexual activity is figured in wiggin's account in relation to heterosex and skin. Yet wiggin may efface safe sex practices in figuring vasectomies as replacements for condoms, even though the associated medical procedures can prevent pregnancy but do not offer any protection against sexually transmitted diseases. Nevertheless, this account hints that there are more masculine iPhone engagements, which according to wiggin are intermeshed with men's sexuality, penetrative sex acts, and digital barebacking and other forms of unprotected sex. Through such narratives, wiggin extends and changes Nye's gender script about the oleophobic coating to more overtly incorporate heterosexual activity.

The Manly Housekeeper's title and banners, which are design elements that often articulate the identity of sites and are thus worth considering, establish his and the whole site's masculine identity. He also maintains a claim to normativity by referencing his wife when explaining oleophobic coatings and "The Proper Way to Clean and Disinfect Your Smartphone."[44] He manages his interest in cleanliness and the ordinarily gender-demarcated housekeeper role by noting that his "wife is a bit of a germaphobe, so she frequently asks" him to "disinfect her iPhone." Through this and related narratives, he renders himself as unconcerned about germs, as a problem solver, and as a form of masculine savior. He enhances his authorial position and prestige by indicating that he is acknowledged by the mainstream press and that he communicated with the *Wall Street Journal* "about the difficulty in cleaning smartphone screens." He also employs the site tagline "Man. Evolved," which positions him as having a more equitable position to gendered labor than other men. Nevertheless, an attention to punctuation, which is also an online close reading strategy, suggests how he employs a period to emphasize and bound his position as a man. In addition, online men's groups often figure men as more evolved and technologically savvy workers as a means of dismissing women's knowledge and labor. For example, the Manly Housekeeper's site banner supports this position by depicting an ape slowly advancing, standing upright, and changing into a man who flips food in a pan. The Manly Housekeeper, in a similar manner to Nye, uses women's bodies to explain the technology and, in this case, to blame them for obsessively desiring clean devices. However, the numerous technology forum posts by men that I study later in this chapter convey their preoccupations with dirty and damaged devices and how they feel.

The popular writing about mobile phones and oleophobic coatings connect the texture of devices to feelings. Nye relates the oleophobic coating to sensations and how a bikini feels when waterlogged.[45] This experience may convey erotic sensations and abjection because of the ways damp clothing feels against people's skin. The journalist Chris Chavez associates oleophobic coatings to pleasurable forms of touching in his instructional article, "Bring back the slick 'new phone' feeling to your display using this amazing wax."[46] He acts as the representative and the company's ideal male user in re-creating the new phone's tactile pleasures, and thereby banishing dirt and unpleasant surfaces. According to Chavez, "There are few things in this world better than taking a brand new smartphone out of the box and sliding your finger across that cold, silky smooth glass. It's an almost sensual experience." As in other texts, Chavez persistently employs the term "feel" and related language. He relates the feeling of fingers on the phone surface to the coating, or skin. In the associated video, Chavez's hands engage in a kind of autonomous sensory meridian response (ASMR) when they constantly touch the screen and rub against each other. His hands act as specialists on touch and as representatives of how, as he rhapsodizes in a kind of postcoital exclamation that "it just feels so good." Chavez's emotive narratives about feeling touchscreen surfaces are also present in Nye and the Manly Housekeeper's discussions.

Popular accounts usually emphasize feeling, but they do not always represent the slick (or textured) surface as delightful. People associate slickness with slipperiness and the fragile nature of the phone, which can be difficult to grasp, slip through fingers, and break upon impact. Rachel Plotnick's research on buttons indicates how individuals reject the "slick, flat, glass" of touchscreens.[47] In a related manner, BigDaddy0790 comments on how his new phone is as "slippery as freaking soap" and has been dropped so many times that "all the edges are roughed up, and mute button gets stuck."[48] dodger_m's "fingers get a little sweaty" when using the phone and the oleophobic screen feels a "little sticky and grainy."[49] These commenters focus on the ways hands feel when in contact with the phone and how individuals may refuse the device's constitution as to-be-touched-ness, including hands that pause at the texture, move away because of dirt, and are not wrapped around and meshed with screens.

Reporting on the Touchscreen and Viral Contamination

People reference women's purported interests in cleanliness when writing about digital devices and the coronavirus pandemic. In a similar manner to the Manly Housekeeper, aldo82's concerns about coronavirus spread and contaminating the "phone and phone case" is buttressed by a reference to the commenter's "wife, who is a doctor" and "has the same apprehensions."[50] Thus, aldo82 moderates individual unease and the risk of a single status by referencing a worried woman. This commenter's position is further elevated by indicating that she is a doctor, and presumably versed in medical issues (although, of course, there are many kinds of doctors, some of which are not medical professionals), and the real distress aldo82 is feeling is for her. aldo82 establishes the couple's physical and emotional relationship and places limits on other types of skin contact, including the Apple leather case's connection to animal and human skin. The Apple leather case's position as skin, as I suggest later in this chapter, is another instance where individuals produce a form of skin-to-skin contact, render layers of flesh, and accelerate concerns about bodily filth.

Reporters and commenters often associate dirty phones with contagion. These concerns have escalated because of the coronavirus. For instance, Paul Czerwinski, who is a director of healthcare sales, offers cleaning and disinfecting methods because "COVID-19 has drawn increased attention to the need to sanitize mobile devices."[51] The journalist David Levine's "How to Clean Your Germy Phone" encourages readers to engage in collaborative forms of disgust and to respond by cleaning.[52] He asserts, "Our phones are really dirty," and while overlooked, they are the "one thing we handle more during the day than perhaps any other object." Levine encourages people to amplify their feelings by Googling "cellphone germs." They will "be grossed out by headlines screaming warnings like Your Cell Phone Is 10 Times Dirtier Than a Toilet Seat (Time magazine), Your Smartphone Screen Is Probably Disgusting (USA Today) and The Dirty Cell Phone: 25,127 Bacteria per Square Inch (StateFoodSafety.com)." Levine thus articulates and remains outside a frantic journalism where writers textually shout until readers' bellies wrench. The screaming journalist is thought to frighten readers into following health protocols *and* to emotionally sicken them. Like Kristeva's narrative about the thin film on top of milk, which causes people to gag, these narratives about biological traces out of place are designed to produce extreme feelings.[53] They script a body that follows journalists' directives and potentially develops more healthful and sanitary practices.

Levine's methods of scripting disgust, in a similar manner to the other practices that I have identified, employs women when referencing filth and demands for cleanliness. His article is illustrated with a photograph of a woman's hands cleaning her phone. In another image, a smiling woman cleans a wood surface. This depiction is linked to a slideshow where women appear to further consent to cleaning, including kneeling to scrub the floor. As Ahmed's consideration of women's hands suggests, these women smile welcomingly and accept directives for their hands to be at the service of others and to fulfill cultural scripts. Some of the women's heads are cropped out of the photographs, as if to suggest that only their laboring hands are necessary. These women are representations of cleaning, and thus the site of dirty devices and hands. Such frameworks, as I continue to suggest in this chapter, use conceptions of filth (and viral contamination) to render women, people of color, individuals in the global South, and other oppressed subjects as servile bodies who are less civilized and human.

The reporters John Harrington and Charles B. Stockdale propose that the coronavirus and fears of contamination have left people to "their own devices at home, and those include cellphones, laptops, iPads, video games and television remote controls."[54] They employ the term "devices" in the kind of ambiguous manner that Barbara Johnson interrogates.[55] In

Harrington and Stockdale's formulation, the pandemic left people to their own somewhat disorganized plans, which include engaging with touch-based devices. The journalists connect the coronavirus to touch and then encourage more managed forms of sociality, contact, and digital device use. State control may also be justified through this analogy. They note that the position of these devices as "high-touch items" and their potential use by many people have increased during the coronavirus pandemic. Touch-screen devices therefore conceptually (and potentially physically) develop a form of ominous shadow or grimy skin—concepts that I discuss in more detail later in this chapter—because their use produces what Harrington and Stockdale describe as the "specter of spreading the dreaded coronavirus." Harrington and Stockdale, as well as other authors, identify viral aspects of touchscreens, which include their mass marketing, adoption as necessary parts of contemporary culture, scripting of cultural mores, and contribution to health risks and contagion. Concerns about touch during the pandemic, as I continue to indicate in the afterword, sometimes downplay viral spread through droplets and aerosols by emphasizing easier-to-manage contaminated surfaces. Nevertheless, a growing number of reporters have identified the focus on cleaning as misdirected and as a form of comforting theater that elides more uncontrollable threats.[56]

Despite greater airborne risks with the coronavirus, people's uncomfortable feelings about contaminated surfaces are emphasized by Juna Xu, who notes, "Sterilising your mobile phone is a smart idea to prevent the spread of COVID-19, but a leading expert has suggested we're not doing it enough."[57] These encouragements to do more cleaning, which can be useful with some illnesses and contaminants, is amplified by titling the article, "Yikes! You should be cleaning your phone more regularly than you think, *takes out antibacterial wipe*." This title is designed to generate shared emotional responses to notions of being polluted. The emotional "Yikes!" with an exclamation mark conveys voices raised in concern. Xu's visceral chronicle of producing an antibacterial wipe to separate the narrator from the device and dirt produces a version of Kristeva's narrative about abjection and wanting to push away other people, bodily products, and polluting things.[58] Xu tries to further prompt people's gut responses and alarm over bodily effluvia when advising that phones are "dirtier than a toilet seat." This vision of a dirty phone and body is tied to women and femininity because the article includes an image of a woman in a pink shirt who is sitting on the toilet and holding a pink phone. While pinkness often evokes normative forms of femininity, her perch on the toilet renders the referenced femininity as more disgusting.

Tali Arbel's article also includes a photograph of a woman using a digital device as a means of representing the "bundle of germs that is your phone."[59] She chides readers, "You should also wash that extension of your hand and breeding ground for germs – your phone," and only then warns them about phone damage. Josh Ocampo references similar cultural aversions, which often do not stop the conveyed practices, by writing that phones are taken "everywhere: Bathrooms, gyms, subways, and buses."[60] Yet Ocampo underscores the associated intimacies that make mobile phones "like an extra limb or appendage." Arbel and Ocampo allude to the ways devices function as cyborgian extensions and bodily parts, as I note earlier in this book. Extra limbs can also be monstrous when they do not meet expectations for physiognomic symmetry and other norms. While they do not all work in this manner, many of these texts represent men as authoritative users who instruct about cleaning practices. Worried women are supposed to do the cleaning, and thoughtless women and flawed subjects are envisioned employing dirty phones. In each of these scenarios, the associated individuals embody gender norms.

Ocampo continues to try to escalate readers' emotional responses to filth when indicating that his "phone is disgusting." The phone becomes a dirty secret (and a record of something like erotic acts). If "you were to flash a blacklight on it, you'd likely find oil, grease stains, and blood" because Ocampo "cracked it six months ago and it's cut" him "several times." The phone is thus portrayed as an intimate clothing article or bodily part, which, like the phone, can be culturally rendered as repulsive. It is also alienated and threatens to "bite back," as Lisa Gitelman and Geoffrey B. Pingree suggest happens with technologies that are thought of with optimism.[61] The mobile phone acts as a kind of stealth actor and wounds him, and it also potentially physically cuts him out of the relationship and emotionally hurts him. The phone's association with the bathroom replaces the usual correlation of digital processing with the mind and thinking. Such frameworks turn the body and phone on and point it away from the head. They relate the mobile phone to such lower bodily acts as wiping the anus, and more generally to the lower bodily stratum, as Mikhail Bakhtin identifies the flipping of cultural norms.[62] The phone is envisioned as an active

agent in its own troubling position, agitating individuals when revealing
the dirty nature of owners, and opening them up to other bodies and surfaces. In such instances, people employ narratives about filthy phones to
regulate what bodies do and what bodies can be. Thus, people's accounts
of mobile phone and other touchscreen devices rely on gender scripts and
health concerns as a means of producing the most binding of norms.

Individuals and the state enforce norms and try to retain varied forms of
autonomy and control, especially at times of uncertainty. People's embodied attachments to digital devices, frustration and disengagement because
of malfunctioning operating systems, and panicked reactions to slippery
and dirty mobile phones can threaten cultural investments in the bounded
body and conceptually and physically stain objects. The related texts and
authors, as my close readings suggest, point to the ways digital practices
advance bodily disgust and incoherence at the same time as they try to
maintain some bodily stability. In many ways, the Internet is part of a gross-
out culture, which is produced through such things as unlabeled pictures
of "goatse" enlarging his anus and revealing the interior of his rectum and
texted penis pics. This gross-out culture could be further addressed through
the literature on abjection, feelings, and attempts to banish some viewers.

Bodily abjection is also linked to device features when zenpoet notes
that the term "oleophobic" makes the poster "think of Olestra chips. Use
the iPhone's new coating, and it causes anal leakage, because the oil on your
fingers no longer has anything to stick to, so it dribbles out your anus."[63]
In a similar manner to the identification of phones as tainted by overuse,
the commenter imagines that the device and coating's bodily refusal can
cause excrement and bodily fluids to become more animate and flow. This
series of skewed cyborgian bodily relations alludes to the literature on the
unhealthy aspects of phones, including the tendency of these devices to
support colonies of bacteria. zenpoet's comments link skin (and the body's
abject effluvia) to the device and evoke a version of Marc Lafrance's understanding of skin as a leaky interface.[64] This theory of a leaky digital interface
can also be employed to address such disparate practices as the release of
private information, hacking, digital cum shots, and spamming. In much of
the popular literature, contaminated devices and the associated leaky interfaces are correlated with ill humans in a manner that escalates the identification of computer hacks and failures as "viruses." This intermeshing
of technology and the body has been emphasized during the coronavirus

pandemic.[65] So too have people's interests in understanding touchscreen-based phones as clean and smooth embodied extensions been confounded by narratives about exposure and abjection.

Feminizing Dirt and Dust in Technology Articles and Patent Literature

A Microsoft at Home article called "How to clean your computer," like many of the texts that I have outlined, associates gender positions with technological hygiene by framing the document with a photograph of a woman. Television and print advertisements for domestic goods, including cleaning products, also represent women working to maintain the home. The author supports cultural constructions of feminine inadequacies and contamination when narrating a "dirty secret" about never cleaning the computer or removing the "crumbs lurking inside" the "keyboard."[66] In adopting this narrative about the "dirty," the text performs some of the same kinds of gender and sexuality scripts as Nye's analogy. This reference to lurking, as well as its association with feminine violability, render women as endangered (and as needing to remain in the purportedly safe home). The text also figures device maintenance as a kind of housekeeping and connects it to women and women's domestic work. The article correlates women with cleaning, but a reference to Jonathon Millman, a chief technology officer of a computer company, stands in for technological authority. He is quoted as noting, "Your computer could fry if you don't keep it clean."[67] The text also confuses the relation between functionality and aesthetics, as these concepts are hierarchized and conflated in many cellphone posters' narratives, when the company provides "five simple steps in the cleanup and maintenance routine" to "keep your computer and accessories looking shiny and new." In many narratives, the gleaming, new-looking phone is correlated with functionality rather than aesthetics. However, as I demonstrate in chapter 1, women are often mocked for choosing what are purported to be aesthetic fingernails rather than dexterous fingers.

Computer and patent literature emphasizes the problems that occur when dust and related contaminating materials get into computers and peripherals. Patrick Bass's 1986 article about the hard disk before the move to solid-state drives relates high speed and the microscopic distance between the reading head and disk with potential risks to the device and encoded information. The author argues that if the "distance between the hard disk and the R/W head was scaled up to one inch, the diameter of a typical human hair would be over 16 feet" and could cause disk damage.[68] Bass scales up connections and thereby amplifies the potential risks and monstrousness of hair and other bodily effluvia. James Stephen Rutledge, Cory Allen Chapman, Kenneth Scott Seethaler, and William Stephen Duncan's patent outline from 2006 similarly notes the risk of forced air-cooling bringing "dust or other particles," which include large amounts of sloughed-off human skin, into the computer enclosure.[69] They warn that these bodily and other residues can produce computer and component failures and fires. Such authors suggest that the body and its parts should be kept away from computer processing. Of course, the body also builds aspects of computers and online systems, codes software and sites, and engages with devices and settings. Keys, buttons, fingerprint readers, hand-pointers, and fingernail grooves reference and establish the link between specific bodies and digital devices.

Keith Evans reminds eHow readers that dust is "composed of dead skin cells, smoke and ash particles, pollen and other natural materials" in "How Does a Computer Monitor Get Dirty?"[70] He notes, "When dust is left uncleaned, as is common on the back and less accessible portions of a monitor, it can permanently stain the monitor's plastic housing." Individuals thus influence monitors through their skin and other bodily traces, which are impressed onto devices and produce their features. In these cases, the body is also beside itself and technological devices, as Nicolette Bragg and others highlight such relationships.[71] Evans notes, "When a person touches the plastic casing of a computer monitor, the natural oils in the person's hand are left behind and can cause dust and dirt to adhere." In these and other cases, the skin of the individual and the accumulated bodily detritus on the computer act as part of the device's leaky interface, which positions and repositions the individual in relation to the technology and expands Lafrance's conceptions of skin.[72] Individuals and computers are also, as evoked by Didier Anzieu's theories, connected through this shared skin.[73] These narratives about computers correlate touching and dirt and indicate that the body makes things dirty.

Annett Davis's Fit Moms Fit Kids Club also suggests how bodies, including children's bodies, produce dirty screens. She furthers the kinds of embodied and gender scripts that I discuss earlier in this book and tries to correlate clean devices with women's ability to meet normative criteria and the household management of dirt and children. She argues, "Cleanliness provides a sense of peace (less stress=better health)," while also narrating her inability to maintain these states.[74] The feminist authors Barbara Ehrenreich and Deirdre English argue that such household technologies are designed for women and should be more carefully interrogated because they represent the "material embodiment of a task, a silent imperative to *work*."[75] There have been debates about the relation between innovations in household technologies and women's time doing domestic work. However, Michael Bittman, James Mahmud Rice, and Judy Wajcman employ data that correlates women's time working in the home with their available appliances to demonstrate that household technologies rarely reduce women's unpaid domestic labor and in some cases they increase such efforts.[76] Ehrenreich, English, and other feminists emphasize that it is impossible for women to meet cultural directives about maintaining perfect households, families, and personal appearances.[77] By titling her site "Fit Moms Fit Kids Club," Davis may remind readers of cultural expectations about women's self-maintenance and caretaking, but the blogger also figures lapses in such control and concerns about contamination. For instance, she identifies how people with children "probably have a ton of finger prints" on their "computer screen. Yucky prints, and nasty kid germs." Davis uses and markets a product that allows Fit Moms Fit Kids Club to reprieve normative gender positions and make computer and phone screens "mirror reflection clean!"

Mirror screen reflections address individuals with images of blurry audiences that look like and are them. They are thus a form of direct address that acknowledges the viewer and a tactile address because this recognition is organized around embodied traces and touching. These references to the mirror and the relationship between caregiver and child evoke Jacques Lacan's and other psychoanalysts' conceptions of the mirror stage.[78] In this stage, the child is expected to begin to acquire a sense of self and notice the agency of the body through a series of identifications and discoveries of differences. Christian Metz employs the mirror stage as a way of analyzing cinema viewers' association with the ideal image on the screen.[79] In his theory of Skin-ego, Anzieu suggests a more messy and embodied series of connections where the body and skin have what Segal describes as a "double surface."[80]

Bodily mirroring and "maintained" screens may work to produce a coherent self by effacing such fragmented traces as body oil, fingerprints, and skin debris. However, the touchscreen-based phone produces multiple and frag-mented versions of the self through user names, avatar images, face and fin-gerprint identifications, app logins, and selfies. With these mobile phones, the individual's skin is wrapped around the screen's skin, illuminated by ambient light and glow, and engaged or disaffected by promises of connectiv-ity. Cultural expectations for clean digital mirror screens also produce a group of women and feminine laborers who watch over and stand in for cultural norms. Ahmed's framework suggests how these women's arms and hands are caught and directed to work at the bequest of cultural expectations. People also dismiss these hands and their fingernails, as I note in chapter 1, for being too embodied and contaminated by dirt from their labor.

Collaboratively Amplifying the "dirt problem" in the MacRumors and Apple Community Forums

People share their intense responses in technology forums when they notice that their mobile phones have bodily matter on or under the clear surface. Skin and hair are often imagined to constitute bodily boundaries, but they generate disgust when found in other people's environments. In many cases, these individuals employ gender scripts to explain their posi-tions. For instance, texasstar1981 foregrounds these feelings about other people's detritus when beginning a MacRumors forum thread with a post titled, "freakin' hair on camera lens - iPhone X."[81] The poster writes, "WTF. just got it 15min ago and immediately spotted a hair under (!) the camera lens." In a similar manner, gbrancante notes how the "freeking Phone has a hair and some dust!"[82] These and other commenters employ curses and intensifying exclamation marks, which I suggest should be noted when closely reading online texts, to convey extreme irritation. They also direct other people to amplify their own feelings.

The post by texasstar1981 focuses people's looks and sentiments by including an image with the hair circled, in red, thereby suggesting that the hair should be bordered off from the owner and the rest of the device (figure 2.1). Given that dirt is what Mary Douglas identifies as matter out of place, texasstar1981's drawn circle is a means of containing such objects and repositioning them in a more acceptable manner.[83] This need to con-trol matter out of place and the body is advanced when predation (whose

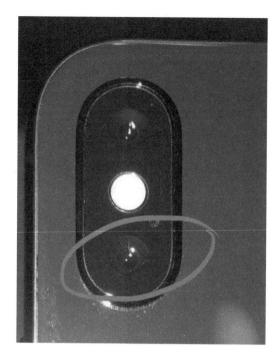

Figure 2.1
Screenshot from texasstar1981, "freakin' hair on camera lens - iPhone X."

member name evokes sexual violation and concerns about such abuses) notes, "At least it's not squiggly hair."[84] This mention of the improperly located pubic hair is a reminder that cultural conceptions of the sexual, especially when referencing the feminine and nonnormative bodies and identities, produce "dirty" bodies and intensified forms of disgust. There are persistent cultural directives to keep a distance from other people's body fluids, especially fluids that, as Kristeva suggests, are deemed more repulsive because they are not clear.[85] People amplify such rules when they express concern about devices being contaminated by fluids and debris and incorporate touchscreen-based phones and other technologies into their notions of self and everyday experiences.

Muriel Dimen describes related forms of excited disgust and shared affective contagion in her article on the "Eew! Factor." She states, "Feelings arrive at once corporeally and psychically."[86] This corporeality "is as much a two-person as a one-body phenomenon." People experience these sensations when their bodies are forced into association with the bodily detritus of unknown individuals. This may happen when people examine texasstar1981's image, read related narratives, and detect more debris in the photograph and other places. Such excitement is also conveyed when martin-2345uk writes, "Another one with a hair?! Does Cousin It work on the production line?"[87] martin2345uk figures a monstrous and nonbinary body contaminating the manufacturing process and the clean and smooth surfaces that people want to associate with their mobile phones. Some facets of cleanliness are also reasserted when texasstar1981 posts an update about getting the "hairy X swapped out" for an "absolutely spotless 'white envelope' unit."[88] In this narrative about improvement, texasstar1981 distinguishes whiteness and spotless packaging and devices from the experiences of production-line workers, dirty hair, and the racialized bodies that are marked as hairy.

Commenters collaboratively produce desires for clean whiteness rather than interrogating the racist aspects of some posts and the experiences of people who produce these technologies. Normative cultural conceptions of body hair, as Elena Frank suggests in her analysis of shaving, elevate white, clean, intelligent, and hairless individuals while denigrating people of color by depicting them as hairy, dirty, lower class, and stupid.[89] These hierarchies provide a structure where fault is found to be with the worker and the worker's bodily matter. This legitimizes corporate control and the drive to efface individuals who manufacture technologies, especially the elision of their bodily labor and calls for recognition, from the devices that they produce. Commenters push workers' hands down, as Ahmed outlines these practices and I consider them in chapter 1, and tie workers' hair up as a means of privileging buyers and the clean, unmarked device. In such narratives, consumers indicate that they want workers' bodies and labor to remain undetectable.

Forum threads about hair and dirt generate affective contagion and upset readers who anticipate similar troubles and imagine themselves in the roles of commenters. If the "phone arrives" with the dirt chronicled by other posters, suggests UL2RA, "it's going to trigger" further "anxiety."[90] In such cases, commenters collaboratively intensify feelings about dirty and otherwise imperfect phones. People's narratives animate the bodily debris and other dirt and intervene in commenters' experiences with devices. For don-ster28, "Just knowing something is moving inside" the device makes the commenter "cringe."[91] Cringing is an embodied response where individuals try to move their bodies corporeally and emotionally away from contaminating elements and people and express extreme affect. Cringing thus suggests a problem for device manufacturers that try to link people to their touchscreens through a variety of design and advertising practices.

In a consideration of cinema as skin, Tarja Laine argues that such bodily experiences as shivering (and I think its relationship to cringing), transform the "skin into a kind of 'resonating membrane.'"[92] She references Steven Connor's indication that when people have an aversion to something, the skin squirms and thus becomes a version of the shifting or disgusting thing that they want to escape.[93] The shared features of skin and the ways people are shaped through contact are foregrounded by these scholars' arguments, including Connor's persistent identification of skin as a membrane. Yet cringing is also an attempt at articulating the bordered body and individuation of the person. Cringing is too often an expression of intolerance that allows people to establish physical and conceptual distance from nonnormative identities and practices, but it can also disrupt the intermeshing of bodies and devices and the associated presumptions that such representations and sensations are unmediated. I employ Laine's and Connor's analyses to indicate the ways individuals and devices resonate with *and against* each other because of responses to the other and dirty devices. This notion of skin as a resonating membrane is also useful in considering, as I do in chapters 3 and 4, the ways emoticons and emoji are designed to get readers to collaboratively feel. It also evokes the ways autonomous meridian sensory response (ASMR) video producers work to facilitate viewers' intense embodied experiences with resonating screens.

The concept of the resonating membrane conjures individuals sharing their concerns about device screens with forum commenters. Irritated commenters offer methods of identifying and responding to what Bleifuss describes as a "dirt problem."[94] These posts about mobile phone problems provide individuals with assurances that they are not alone, a community to fret with, and information about diagnoses and solutions. Some individuals also refuse to be irritated by problems and instead express annoyance with posters who itemize device deficiencies. For example, some posters negatively responded to rmoliv's concern that the new iPhone "got scratched all over."[95] Starship67 argues, "Most people cant be bothered to look that hard to find something wrong."[96] MacDawg "can't even be bothered to turn" the "phone over to look."[97] Such commenters convey indifference and turn toward and away from the problem object and poster. Their dismissive responses are in some cases designed to bait and intensify the bother experienced by concerned commenters. To insist on being unbothered is to place pressure on other individuals to accept their position, even if it is troubling or unlivable. Such commenters convey indifference as a means of producing the kinds of emotional baiting that Sarah Sobieraj and Jeffrey M. Berry analyze and that I consider in more detail in chapter 1.[98] While the posts that I have outlined are designed to further incite participants, they do not challenge the cultural structures and hierarchical identities produced in the original posts.

Most of the posts that I consider in this chapter are not feminist. Some of them contain anti-feminist and misogynistic conceptions in their addresses to "guys"; in their insistence that women function as erotic objects; in their association of dirt with women, the feminine, and people of color; and in their indications that women have delimited relationships with technologies. However, I employ Ahmed's analysis to consider the ways contemporary culture displaces critics and critiques, as well as how it dismisses feminisms. Ahmed notes, "We learn about the feminist cause by the bother feminism causes; by how feminism comes up in public culture as a site of disturbance."[99] Thus, as I suggest in chapter 1, feminist politics is understood and encountered by anti-feminist culture and society more broadly as disruptive. The "bothered" forum commenter is also made into a bother and a problem that disrupts the comfortable flow of technological and consumer pleasure. People self-present as unaffected as a means of preventing other individuals from expressing their dissatisfaction. Purportedly unbothered people switch the problem and the bother from the device to the individual. Nevertheless, these critiques receive less vitriol than women's challenges to gendered iPhone scripts.

Scratching, Irritating, and Feeling Fingernails

People's queries about greasy and smudged screens are reminders of the ways bodies (and phones) act as leaky interfaces, including the ways they combine with bodies and leak personal data. Remington Steel proposes that a poster who asks about fingerprints has "greasy finger syndrome," as a means of creating a category of abject people and separating them from normative masculine bodies and clean screens.[100] Newtons Apple dismissively constitutes MacKid1983 as an obsessive and filthy nerd for asking, "Anyone experience really bad smudging on your iPhone 7 display screen?"[101] MacKid1983 also notes that the "fingerprints are ridiculous" and wonders about "bringing it to apple and complaining." Newtons Apple blames MacKid1983, suggesting, "Clean the screen once a day and try not to use it while you are eating fried chicken."[102] While the nerd has historically been characterized as immature, engaged in poor eating habits, and unconcerned about "his" body, as I suggest in *The Body and the Screen*, the more recent mainstreaming of computer culture has resulted in the elevation of the associated individuals and conceptual, if not physical, reshaping of their bodies.[103] Remington Steel and Newtons Apple engage these shifted frameworks and try to elevate their positions by articulating, and then distancing themselves from, abject bodily traces and the associated people.

Apple's marketing of iPhones, as I suggest in chapter 1, identifies fingers as the ideal interface tool for touchscreens. However, Remington Steel and Newtons Apple indicate instances where the actions of fingers on screens cause problems for people who are using touchscreens, as well as for the people who rely on aspects of screens for their identities. People's negative comments about greasy phones suggest that bodies must remain nonporous and bordered so that they match the requirements of devices. People also identify fingernails, which evoke women, femininity, and the nonhuman, as problems when engaging with the capacitive touchscreen interface, as I note in chapter 1 and in this section. The nail evokes feelings because it is hard in a way that other parts of fingers are not.

People express concerns about fingernails touching computers and capacitive touchscreens. For instance, johnny_240sx is concerned because a friend's "nail made contact."[104] johnny_240sx provides a kind of schematic map of personal and technological risk wherein fingernails and their feminine connotations are less threatening because they "weren't long and they were plain (it wasn't painted it had no fake crap on it just straight up vanilla)." The poster also wonders, could they "create a scratch on the display" when someone uses a "fingernail to point" and rub "it across the screen?" CCato77's visceral response is an expletive like "******."[105] While individuals are understandably concerned about potential damage to their screens, fingernails do not ordinarily pose such risks. Fingernails are less likely to smudge screens than are greasy fingertips. Yet, like the willful, raised, and insistent arm that Ahmed references and I theorize about in chapter 1, the feminine and pointing fingernail with polish is refused and pushed down by posters' narratives about contamination.

Individuals indicate that the small physical influence of fingernails has intense emotional impacts. Lika_tm proposes that a fingernail made a "tiny scratch on the touchpad."[106] The commenter questions "why a touchpad should be so easily damageable if it requires constant touching." In writing this, Lika_tm offers a concept of touching that incorporates and resists fingernails. The poster also appears to temper the negative gender connotations produced by fingernails by insisting that the nail is "not a very sharp one." These comments may continue the identification of women, and the fingernails that they are thought to select "instead" of technological proficiency, as unintelligent. A large number of technology forum posters respond negatively to longer fingernails and ornate nail art, as the previous comments indicate. These posters correlate such embodied and aesthetic choices with unappealing forms of femininity, race, sexuality, and class. Of course, the varied sites where people share and demonstrate nail polish applications tend to promote positive reactions to fingernails, nail polish, and nail art. Nevertheless, even in nail art forums, which I have considered at length in previous research, individuals articulate limits on what is deemed to be acceptable forms of nail length, decoration, and femininity.[107]

Fingernails, as I have begun to suggest, are feeling devices and tools. Yet their work on screens also causes concern and unease. For instance, jazzdude9792 describes being "pretty happy" with his iPod until noticing a "small scratch in the glass screen. This isn't any ordinary scratch," argues jazzdude9792; he could feel it with his "finger nail pretty easily meaning that it is a substantial gouge in the screen."[108] The "grooves" in alyabiev's screen "can be felt not just by a nail, but with a naked fingertip."[109] These posters' pleasure is disrupted by the investigative abilities of nails and fingertips, which they employ even while knowing what they will feel and find. On such occasions, the fingernail is hooked into and feels the scratch, which causes the nail and the individual's attention to the larger surface to stop. Zach the Apple User's "distracting" iPod scratch can be felt when running a "finger nail across it."[110] Individuals use their fingernails to detect changes in surface and texture, but these instances of feeling nails, which sense surfaces and convey physical and emotional information about fingernails and other things, also cause commenters to be uneasy and trigger unpleasant feelings in other people. As Dimen suggests, such affective experiences are collaborative.[111]

People's concerns about scratches and the many incidents that produce them are amplified when kaans warns that the iPhone X "screen literally gets carved just by random dust, it's not something you can ignore either, your fingers/nail will catch onto it."[112] Fingernails thus function as useful tools and discover and know the screen, surface, and individuals' feelings. Fingernails recognize scratches in a manner that seems to shift knowledge processing and storage from the brain to the hand. The nail functions in a similar manner to Vivian Sobchack's hands and emotionally feels and physically recognizes things.[113] Narratives about screen scratches tend to emphasize the experience of feeling the screen surface by touching rather than seeing it. This happens when WISG.1 pushes on the "iPhone screen (5C)" and it "moves in and out."[114] Individuals press and pry at seams in the smooth surfaces of devices. Anne Cranny-Francis describes a related physical and narrative "seamfulness" and semiotic "semefulness" where individuals attend to multiple meanings, "seams," and how technologies are sutured into everyday life.[115] According to Cranny-Francis, the identification of the "seamful interface is also 'semeful' in that it draws the attention of users to the interface and hence to the ways in which it 'makes meanings.'" Digital scratches and seams are thought to become a kind of bad writing and meaning production. However, they direct individuals to focus on how devices are conceptualized and produced.

Nanna Verhoeff proposes the concept of the theoretical console to explain individuals' employment of the dual screens of the Nintendo DS. She argues that the interface encourages people to think about how multiple forms of engagement and representation are enabled. As Verhoeff suggests, such devices act as theoretical objects and consoles because they raise "questions about the specificity of the screen gadget as object, and about the entanglement of technologies, applications, and practices."[116] The scratch does similar work in getting people to conceptually engage multiple registers of the device, including as a screen object and a technological apparatus. These mars link and shift individuals between finger digits and the digital device. When individuals use their fingers to pick at flaws and seams that they fear will open and spread, the sensed scratches generate affective experiences that are related to individuals' apprehensions about dirt in devices.

Such mars function as a form of punctum, which I elaborate on in the introduction and chapter 3, that shocks and wounds device owners when they
unexpectedly discover damage. While Roland Barthes's theory of punctum
addresses photographs and is at least initially associated with visual objects
that produce feelings, these touchscreen wounds are experienced through
hands as well as eyes and shift between haptic and optic experiences.[117]

Cracked Screens

Individuals' hands and fingernails as well as eyes experience tactile surfaces and pause at scratches and other abrasions when employing touchscreens. Bodies may be mirrored and marked on devices, as I suggest earlier
in this chapter and in chapter 4. In *The Body and the Screen*, I consider how
the positioning of women webcam operators as objects is torqued by their
control of the technology and the ways viewers are reflected on screens
and conflated with screen images.[118] Viewers thus watch versions of themselves. Touchscreens broadcast material traces of individuals to owners and
onlookers and provide fingers and nails with unwanted information about
damage. Dirty and damaged screens frustrate and otherwise emotionally
influence individuals by failing to deliver on corporate promises of new and
"self-cleaning" oleophobic surfaces. People also associate cracked screens,
which are often viewed as dirty because they gather bodily and other matter in their crevices, with emotional sentiments.

People regularly reference emotional reactions to phone cracks as means
of establishing differences in age, class, gender, race, and sexuality. In many
accounts about women's broken screens, men distinguish between men's
responsibility and women's technological disinterest. For instance, nasa25
notes that his girlfriend's "screen already looks like its been through a gulf
war."[119] Another commenter, kayzee, indicates that his girlfriend has "completely destroyed her Z3 Compact :disappointed: cracked both the front
and rear."[120] These people figure themselves and other men as judges of
technologically acceptable behavior and as the more adult and knowing
employers of devices. Thus, men who break or worry over their phones
are often characterized as childish in a manner that sustains patriarchal
structures where older men are privileged. However, men's "girlfriends"
and other women who are romantic partners are still considered with
increased scrutiny. While ThatiPhoneKid, who self-represents as a juvenile

brand enthusiast through his member name, narrates multiple technological disasters, forum participants are particularly interested in his concerns that his girlfriend's "hoovering" over his phone produced "internal damage."[121] Commenters correlate "hoovering" to oral sex and represent women as capacious erotic objects. Women's adequacy is further questioned when Relentless Power asks, "How did your girlfriend end up vacuuming over your iPhone without seeing it?" ThatiPhoneKid replies, "shes reckless lol." Commenters use the post as a means of reasserting traditional women's roles and produce the kinds of gender positions that I analyze in this and other chapters. For instance, keysofanxiety suggests that women have limitations when writing, "at least your missus does some housework."[122] keysofanxiety thereby encourages the misogynistic idea that women should do household labor and disappointment that they inadequately perform this work.

HappyDude20 wonders about and reproduces class structures when asking, "Is a Broken iPhone Screen Unprofessional/Embarrassing" and "do you think it looks bad?"[123] The commenter inquires about "bad" appearances and connects professionalism to aesthetic norms and economic attainment, such as being able to pay for new phones. Other participants, including scaredpoet, respond using the same terms. For instance, scaredpoet feels "bad for them, both for having such bad luck and for not being able to get it fixed."[124] HappyDude20 is warned by scaredpoet not to expect other people to "use your busted phone and risk cutting" their "fingers." Such warnings suggest that cracked screens *and* dirty devices and bodies produce abject experiences and keep other people at a distance. Furthermore, scaredpoet, whose user name expresses apprehension, indicates that the skin and fingers of viewers respond to individuals with broken screens and conceptually move away from any contact.

HappyDude20 asks about different factors that might result in a person with a cracked screen looking unprofessional. His digital self-presentation and query are also tempered by his member name, which evokes the relaxed and unambitious guy. However, many individuals who reply correlate touchscreen-based phones with white-collar professionals. JohnLT13 asserts, "If one can afford an iPhone, you would figure they could afford to get it fixed" and "Broken screens look shabby."[125] This associates new screens with respectability and financial success and broken screens with poor economic decisions and dilapidated environments. This indication that screen aesthetics (and possibly functionality) influence people's status is also established when babycake describes a school administrator with a cracked screen, "who's reprimanding an ambassador's kid."[126] According to babycake's scenario, the ambassador meets with and sees the administrator's cracked screen, thinks "how trashy, can't even afford to fix his phone," and identifies "how ghetto your school is." In using a version of the phrase "that is so ghetto," babycake continues the unfortunate association of economic and educational disparity with violent crime and blackness.[127] Early uses of the term "ghetto" referred to areas where Jewish people were confined. It now describes an extremely populous slum, which is occupied by minority groups who are situated there because of economic and social factors. As Kenzo K. Sung suggests, describing something as "ghetto" is ordinarily disparaging. It essentializes the "structured oppression of impoverished and racially segregated communities as pathologically inevitable and isolated from 'mainstream' society."[128] Individuals are blamed for living in poverty in a ghetto and believed to behave in a criminal and indolent manner that explains and justifies such habitation.

Class positions are further established *and* babycake self-identifies as distinct from poor individuals with broken phones when instructing readers, "When you get into a high power position, looks matter; you have to tie your tie right, wear the right glasses, have the right type of skin, and have a nice phone." The reference to skin connects individuals employing phones to the culturally acceptable surfaces of bodies and devices. babycake correlates power with norms and suggests that individuals should follow cultural proscriptions. The poster recommends the "right type of skin" and risks making racial judgments that suggest that whiteness should be associated with authority. If "ghetto," as it ordinarily does in contemporary narratives, is intolerantly associated with black viciousness, depravity, and bad taste, then good skin proposes the opposite: an attractive and enabled white subject. Through such accounts, skin is rendered as an attribute and component of the individual that can be prosthetically donned and removed in the same way as well-styled ties, new and pristine mobile devices, and other accoutrements are purportedly under the control of every individual. This elides systemic racism and other forms of disenfranchisement and renders people's positions as personally achieved or their fault. People employ the mobile phone and skin in such narratives as a form of theoretical console, but these elements are largely referenced to reinforce hierarchies rather than as means of intervening in cultural structures.

Phone Case Patina and Skin

Commenters mention skin in their writing about smudged and dirty phones and in their considerations of phone cases. For instance, as I suggest earlier in this chapter, iphonefreak450 describes such screen buildup as fingerprints, smears, and "film."[129] BugeyeSTI chronicles "clear skins that can be applied to the back of the phone" to prevent scratching.[130] Individuals also associate embodied and device skins in posts about Apple's leather cases, which were once a part of an animal. In general terms, phone cases are records and extensions of skin and of individuals' actions. Commenters establish these connections in posts where they describe the tactile aspects of leather cases and the ways their embodied engagements influence the color and feel of the case. For instance, a thread about "Apple Leather Cases - Patina Proud Photos" convinces sean000 to get a saddle brown case "because it looks so great as it darkens over time."[131] In such threads, leather case patina is ordinarily identified as the changing color and texture of the case.

Some people insist that patina is beautiful and others assert that the production of patina through bodily excretions is disgusting. In each case, they focus on the kinds of aesthetic concerns about embodiment and connections to phones that women are condemned for. Posters frame their positions in such threads as "Which Apple leather is best for not showing dirt/patina?" As the original poster of this thread, MrMister111 indicates that he owned an "Apple leather Red" but "must have greasy hands as it wore bad, and always looked dirty."[132] The poster suggests that the case functions as a kind of theoretical console and informs him about and makes him recognize the filthiness of his body and the characteristic of his skin. In a related manner, Ralfi is "not a fan of the 'Patina' effect" because "it's mainly a build up of body oils/gunk on your case."[133]

By expressing dislike for what they identify as the residual sweat and accumulations of owners' bodies, MrMister111, Ralfi, and other individuals foreground the sticky ties that bind phones, cases, bodily residue, and individuals together. Whether classified as an ideal or unappealing conjunction, posters make decisions based on these connections. For instance, Ralfi chose "Silicone instead of leather" because the "weekly wiping down with a damp cloth will remove the sweat, dirt, lint etc. in one *foul* swoop."[134] This represents cleaning, like the use of leather cases, as a disgusting activity and something that purifies surfaces that are not leather. The condition of the leather case "grosses" Ralfi out because the commenter keeps "thinking this Patina effect is just a combination of the absorption of human oils, grime & wear on the case. Does the case start to smell at all after a while?"[135] This poster provides details about bodily effluvia as a means of establishing a distance from the associated mobile bodies, which mesh embodied traces and mobile devices, and the material feel and content of cases. Ralfi and other commenters who resist leather cases also enact a version of excited disgust, which Dimen identifies, and thereby they reanimate the cases even as they try to keep the associated notions of contamination away from their bodies.[136]

Ralfi argues that new leather cases "look brilliant, but then they change into something that's not very attractive." Cases are figured as unappealingly morphing and transforming. Such unstable objects and bodies may cause unease in situations where individuals and cultures are already being reconceptualized by varied social and political movements. Ralfi wants to keep at a distance from the material shifts of these mobile device conglomerations and "wouldn't want to touch a typical owners leather case after they've had it a while - like shaking hands with someone who hasn't washed in months." Ralfi renders the case as a stand-in for and corporeal version of the owner, and as part of a great unwashed culture (and phones), which evokes babycake's rendering of the ghetto. As such comments and the earlier posts that I outline suggest, people construct the status of and feelings about iPhone and other touchscreen-based phone owners in their narratives about the cost, condition, and usability of devices.

Some commenters reject the functions of phone cases and devices as points of contact. chriscrowlee echoes the aforementioned desires to separate from other bodies when writing, "Leather is skin...and even when treated is a bacteria magnet."[137] As the commenter suggests, leather is skin and thus has a material and structural relationship to individuals' bodies and hands. Anzieu describes how multiple layers of skin wrapping facilitate interfaces and communication between people and things.[138] Skin is a layer and a wrapping that covers and protects devices, connects individuals to technologies, and manages owners' feelings about devices. Ralfi, chriscrowlee, and other posters' resistance to material changes, which is also a resistance to certain cultural notions of movement, is not consonant with the notion of mobile phones. Yet they advance the idea that cases and devices extend owners' bodies toward repulsed individuals by buying and encouraging a different relationship. Fingernails and their links to women and people of color, as I suggest in chapters 1 and 4, are also identified as horrifically moving toward other individuals, extending through the interface, and turning into animal and monstrous claws. Thus, device films and fingernails are useful for scripting normative ends and should be further read and theorized as a means of considering what these practices produce and how they might be changed into a form of Ahmed's resistant arms.

The leather case, in a similar manner as the mobile phone, is animated and attached to the owner. It wraps the owner (and the phone) in a desired or an objectionable additional skin. Thus, Anzieu's and other scholars' theories of skin provide methods for understanding how individuals, devices, and cases become incorporated. Feminist theory, including Bragg's writing about mother-child bonds, also foregrounds the relative and amorphous nature of the individual body and how it is shaped when enmeshed with other forms.[139] Ryan1524 proposes the ways objects combine and that it is "almost as if the device, as a thing, is more complete with the leather cases."[140] The case is also incorporated into the device in a manner that continues its wrapping and covering of some of the phone's surfaces. Ryan1524 considers the "phone with an Apple Leather case as the more complete product now. Naked feels...naked. Anything else feels and looks wrong." This poster picks up on Apple's marketing of accessories and peripherals and becomes a brand supporter. The associated comments figure the leather case as a skin wrapper, as clothing, and as less naked (and possibly less embodied) than the device. Ryan1524 identifies the "naked" phone, which is how people describe the employment of mobile devices without cases, as being incomplete. Yet many positive narratives about naked iPhones, as I suggest next, portray uncased phones as the ideal versions. In these instances, the naked device is thought to better connect the individual to the mobile phone and digital embodiment and to provide erotic experiences. As with other structurations of mobile phones, the naked phone is also correlated with women and the feminine.

Screen Protectors and Going Naked

Molly McHugh reports that the "call to ditch the iPhone case has been around as long as the iPhone has" and "it's a style choice."[141] Her article emphasizes consumers' investments in the aesthetic features of iPhones, but she also wonders why a device that has been around in varied iterations for many years is not accepted if it has "scars. Why is something that is increasingly ordinary also deemed increasingly precious?" McHugh's articulation of the scarred phone is similar to other individuals' figurations of the phone as a version of the individual and as a skin. While she suggests that individuals should be less concerned about phone aesthetics and embrace the device without a case, many commenters espouse the phone without the case as a way of celebrating its design features. The associated transference of the individual's skin to the phone and its surfaces results in referring to the uncovered phone as "naked." In addition, the user of the phone is identified as male, but the phone itself is usually explained through women's bodies, as I have already outlined.

People's normative conceptions of nudity and nakedness are ordinarily linked to binary gender and sexuality categories. Art historians have noted the historical tendency to produce paintings and other visual versions of the female nude.[142] The underlying canvases, like iPhones and their features, are often explained by referencing women's bodies.[143] This is related to the continued eroticization of women, including their constitution as to-be-looked-at-ness and to-be-touched-ness, and their association with the natural (and contrarily artifice and the artificial). NBAasDOGG figures the naked iPhone as a form of visual, haptic, and erotic pleasure when telling "you a story" that is also a gender script.[144] NBAasDOGG "saw a girl" with "her (naked) iPhone X" and thought "WOW, that's a gorgeous phone when you see it in someone else's hands." The author notes that the device is scratched but concludes, "she was gorgeous and she was enjoying the heck out of her gorgeous phone. Maybe we should do that as well." In NBAasDOGG's story, it is unclear if the narrator and readers are supposed to enjoy the woman, her phone, or both.

The narrator here and in too many other frameworks controls the "girl." Jessalynn Keller argues that girling supports dominant culture and tries to keep girls and women from participating in the political sphere.[145] In the case of NBAasDOGG's narrative, the girl is employed as a means of tantalizing and instructing readers. She remains not fully formed because of the poster's employment of the diminishing term "girl" and because of the ways she melts into the device, heterosexual male viewers' desires, and the moralizing account. The woman's attractive body is thus a script and a lens that focuses and reflects male viewers' interests in the phone and constitutes the woman and device as to-be-touched-ness. Even more than the digital phone, she is a story and a device that allow him to make a point. The digital scholar Sarah Murray addresses such practices and, in a similar manner as gender script theorists, encourages scholars to consider "which bodies find themselves with a recognizable fit" when engaging with popular culture and technologies.[146] In NBAasDOGG's and other accounts, heterosexual males are coupled with and enabled to fit the gendered device. I also note, and suggest scholars should consider, the bodies (such as the "girl") that are imagined as mere controllable devices and being directed to fit other individuals. NBAasDOGG employs the figure of the girl in the manner that Ahmed critiques, and also encourages men to imagine women as (and in) their hands and as their erotic laborers.[147]

A similar gender script is employed by davidec to encourage people to use the phone without a case. davidec states that if he is buying a "technological artwork," then he is "going to enjoy it in its naked beauty." When he married his wife, he "didn't cryogenically freeze her to stop her showing the signs of ageing."[148] It may seem as though davidec is accepting women's changing appearance, which of course would be consonant with his own aging. However, he does this in order to make women into tools and examples. Men's processes of embodied change remain unaddressed in his narrative, so that men can nobly accept imperfect women while remaining rational thinkers rather than bodies. In a comparable manner to Bill Nye, davidec and NBAasDOGG use women's bodies as methods of explaining their relationship to technologies. Through such texts, men script the gender and sexual position of the ideal user as male and heterosexual. Male posters' assertions of being sexually interested in women and being experts on women's bodies and experiences may thus work as a form of "no homo," which protects men from and connects them in queer relations with each other and the iPhones that they admire and desire. The phrase "no homo" evokes gay relationality, including men's desires for and associations with digital devices, where such erotics have not been previously highlighted. Men's use of "no homo" excessively, and thus queerly, denies such attachments.

When iwonder36 references the naked phone and notes, "So few things can now be enjoyed without protection," he asserts an active and unimpeded sexuality, which is shared mostly among men.[149] The commenter associates the naked phone with sexual liberation that is not constrained by concerns about sexually transmitted diseases, unwanted pregnancies, or

consent. In a related manner, wiggin, as I note earlier in this chapter, relates
the decision to use a screen protector to the purportedly excessive choice to
get a vasectomy and then use a condom.[150] These posters celebrate an active
sexuality that tends to be withheld from or used to denigrate women. The
phone is further equated to sex when bigjnyc titles a post, "One month
later...anyone else's iPhone X still a virgin?"[151] This phone too becomes the
fixture of the owner's interests and an available, but also dismissed, erotic
object. The original poster looks at the lightning port and realizes that
"nothings ever been in there." This narrative about penetration is echoed
by Vermifuge, who "jammed the dongle up in there day one" and understands the cellphone as a passive and feminine receptacle.[152] Such texts figure a form of heterosex where male penetration and control are imagined
as the qualifications for sexual activity and device use. The employment of
terms like "jammed" and "violated" renders aggressively male gender and
sexual scripts where the pleasures and consent of women or bottoms are
displaced.[153] These dismissals of consent, as I suggest in the afterword, are
escalated because of people's mourning of close contact during the coronavirus pandemic and the overblown claim that #MeToo activism is an earlier
instance of, and at fault for, contemporary curtailments of touching.

Conclusion: Naked Devices and Feeling Hearts

People's concepts of naked phones, like their narratives about touchscreens,
convey ways of feeling, including aesthetic admiration for devices and tactile pleasures in experiencing material objects. For instance, haqsha23 notes
that the "silver version is just so damn beautiful and feels so good in your
hand."[154] In such instances, feeling is a cultural narrative that allows people
to claim their individuality and difference when engaging with massified
objects. The individual object is supposed to be personally felt and held in
a specific hand, which, as Murray suggests, is organized around particular
embodied and identity fits.[155] Yet feelings, as this and other chapters in this
book indicate, can convey company frameworks through personal narratives that erase or ignore underlying cultural strictures. The term "feeling"
is supposed to designate very particular engagements while pointing to an
array of physical sensations and emotional sentiments. It also underscores
corporeal connections and culturally shared moods, which in the cases that
I consider are technologically facilitated and mediated but often identified

as being without intervention. When experienced as abjection, sensations distance individuals from things that make them feel badly or feel too much. Thus, feelings allow individuals to move away from the very elements that trigger and render feelings while maintaining cultural standards.

Touchscreens can be understood and theorized by attending to how these devices are associated with skin and bodies. This includes the ways touchscreens are in contact with, enwrapped in, maintain residues of, and are identified as versions of skin. The features of this skin are also designed to suggest how people should feel about other bodies and devices. Researching the relationship between devices and skin thus illustrates the deeply produced and scripted features of digital experiences and devices. Such research can also be employed to consider how embodied skin functions and the ways it is constructed and technologized. Skin and body studies do not ordinarily address the ways digital devices are connected to embodied individuals and the ways they are identified as skin. Nevertheless, scholarship on embodiment, identity, and new media are advanced and complicated, as I suggest in this chapter, by recognizing the persistent association of skin and touchscreens. This includes how such connections produce gender and other scripts. Feminists can also further intervene in traditional identities and embodied norms by highlighting the functions of and critiquing such digital formulations. Critical interventions into the ways devices structure bodies and script expected users become increasingly important as people further incorporate mobile touchscreens and other screen-based digital devices into their everyday lives.

Some people refuse to interrogate the ways devices are scripted, as suggested by individuals' dismissals of Erica Watson-Currie's critique of the affordances of iPhones, which I analyze in chapter 1. The normative commenters that I study throughout this book enforce ideas about unmediated and intimate connections between bodies and touchscreens by expressing love for their devices and the brand. Love is culturally identified as one of the more powerful and elusive forms of feeling. Love and hearting are also insistently represented through online communication and interface options, including Facebook's heart reaction and Twitter's heart that conveys a like. People's love can also be turned into indifference and disgust. This includes the kinds of revulsion that I consider in this chapter, where intuitive tactile experiences are interrupted or experienced as too much. For instance, individuals who notice their damaged or otherwise compromised screens often express their disengagement because they conceive of device use as intimate and unmediated. As I indicate in the introduction, Apple promotes the idea of intimate engagement, including the notion that devices provide access to loved things. People's investments in and companies' consumer interests in unmediated feelings, and the associated risks of engaging in sentiments such as hearting that are culturally identified as low and feminine, suggest why online representations of hearts and love have been understudied. This critical displacement of the structuring of feeling, which too often produces normative identity scripts, is also why I advocate for such analysis in this chapter and the following chapters.

3 The "heart of social media": Configuring Love Buttons, Hearting, and Members' Gender and Feelings

My email client delivers lists of messages that are punctuated by hearts.[1] These hearts accent titles and replace words in messages. When these emails are part of marketing messages, I am supposed to read the associated hearts as the love that brands and products have for me, the consumer. They also model the kinds of love and enthusiasm that I am expected to convey in return. The tendency to render these hearts as red magnifies the associated narratives about intense emotions and evokes the familiar and scripted sentiments of Valentine's Day. Such heart icons directly address and engage me as an individual. They constitute me and other readers as feeling buyers and friends, as individuals who are in some form of relationship with the sender, and as people who will respond with heartfelt sentiments. The producers of such messages and their heart icons thus constitute feeling individuals and address these readers as some version of "you." By design, they incorporate intensity and care into settings and interfaces. Texted heart emoji, heart buttons, and other online hearts punctuate Internet engagements with feelings. They speckle the digital with programmed and bodily intensities and assert that corporeal individuals are emotionally experiencing these texts while they are in front of screens. Thus, hearts render touchscreen experiences that are about being emotionally touched.

Online representations of hearts widely circulate. The Global Language Monitor listed the heart and love emoji as the top word in 2014 because of its prevalence.[2] Journalists reported on the *Oxford English Dictionary*'s (OED) addition of the heart emoji in 2011.[3] The OED normalized the heart emoji and shaped it to fit the terms of the dictionary by including the listing in its textual form under the term "heart." The OED also identifies the colloquial function of the heart icon, which it associates with insignia and logos that employ the "symbol of a heart to denote the verb 'love.'"[4] Even with the OED's historicization and textualization of the heart symbol, many reporters reacted negatively to it being listed in a respected dictionary and to its more general connotations. For example, Alexandra Petri identifies the dictionary's inclusion as "embarrassing" and as being similar to the purportedly age-inappropriate behavior of your "grandmother wearing glittery makeup and jeggings."[5] Petri also engages in ageist and misogynistic judgments in associating the heart with childish and feminine things. Such responses are echoed in the binary structures and evaluative sentiments that I analyze throughout this chapter. I consider people's and sites' correlations of heart and love throughout this chapter. I also note the kinds of bodies and relationships that are associated with hearts and love. For instance, the term "love" is linked to familial and heterosexual bonds. Many social networks, especially those that address women, deploy heart icons as a means of evoking heterosexual unions, femininity, and loving feelings that tie individuals to sites. Numerous individuals also reject hearts because they identify them as too feminine, childish, and gay.

The heart is part of an elaborate series of new media mood emoticons, emoji, and expressions, which also include angry and crying faces. Social networks use heart and other selectable buttons to structure participants, relationships, sites, communities, and brand attachments. Photo-sharing sites like Pinterest and Instagram and social commerce sites like Etsy and Modcloth offer coded buttons with heart representations that encourage participants to express loving and heartfelt feelings when favoriting people, goods, and content. Social networks also incorporate coded features and narratives about love, heart, and hearting into their self-conceptions. As a method of addressing online representations of hearts and associated conceptions of love, I outline some of the popular uses and definitions of hearts and then analyze the ways Etsy, Facebook, and Twitter employ them. I argue that hearts are used to configure individuals and amplify their sentiments. This configuration includes the association of women with hearts and the implied loving and emotional sentiments. Yet the digital correlation of women and femininity with hearts and love is not inherently productive for the referenced subjects in all settings. While hearting is a useful way of connecting with some consumers and asserting brand identities, these practices also diminish women because they are identified as low and contaminating. Scholars should study such representations of hearting and loving because they have a significant influence on digital engagements and the ways online identities are understood.

Online sites, especially settings that address women, often reference and represent hearts. They also list a variety of elements as the "heart of social media." Site designers and reporters associate what they describe as the "heart of social media" with such things as authenticity (a concept that I consider in chapter 4), blogs, customers, relationships, and sharing. For instance, Jonny Evans reports that people's ability to "publish their own content" and purportedly be empowered by individual expressions is the "heart of social media."[6] Other people relate a charming video of a rounded cat meowing while being exercised and other humorous and uplifting representations of people and animals with capturing the "heart of social media."[7] Such posts generate hearts in the form of Facebook and other reactions.[8] They also work to pinpoint and thereby produce a center or centers and an affective network of online communication. Social network sites and technology companies, in a similar manner to the hearts in emails, use the term "heart" to connect people to interfaces, products, and feelings. Thus, people feel with and for the cat and, according to Apple, locate the heart in technologies. Apple claims that at the "heart of iPad Pro lies the new A9X, our third-generation chip."[9] Through such texts, the company emotionally animates technologies and links them to human functioning, as I suggest in more detail in the introduction and other parts of this book. Such emotive links between animate and inanimate things are purported to provide people with physical connections.

Etsy and other ecommerce sites often connect participants' bodies and feelings to online settings. For example, Etsy's "About" page asserts that the "heart and soul" of the site is its "global community."[10] It uses the phrase to evoke and intermesh the hearts of readers, the essence and feelings of the site, and passionate community. Etsy ordinarily has content on the front part of its site that references impassioned feelings, familial connections, and consumer sentiments. In one such panel, viewers are encouraged to "Find your perfect 'I do' details. Shop wedding." A photograph of the back of a bride's head appointed with a flower crown accompanies the text. The advertisement and associated link relate love and marriage commitments to an additional "I do" of purchasing wedding elements, and correlate the site with intense consumer feelings and physical events. Etsy addresses everyone with directives to buy and articulates the consumer reply in the form of an "I do" that "marries" individuals to the site. By depicting the back of the bride's head, Etsy further renders viewers who are looking at the site as consenting shoppers. Yet Etsy also depicts the woman buyer as not everyone because the back of her neck is visibly light-skinned. The site tends to configure its sellers and buyers (who are also demarcated as crafters and consumers), as I suggest in previous literature and later in this chapter, as white heterosexual women.[11] This is somewhat different from, but can also work with, the usual gender scripts and association of programming and technological aptitude with white heterosexual men.

In this chapter, I connect the scholarship on gender scripts, which I have outlined and elaborated upon in chapter 1, to Steve Woolgar's related research on "configuring the user."[12] I employ the literature on brand love and community and the continental philosopher Roland Barthes's theory of photographic punctum and intense bodily sensations to consider the digital production of sentiments.[13] I demonstrate how these texts provide critical frameworks for considering the ways sites employ hearts and associate them with particular forms of love, relationality, and identity. I continue my considerations of gender norms from previous chapters and suggest how culture correlates hearts and love with women and girls, femininity, queerness, excessiveness, and intense relationships. These associations influence people's employment of heart icons, reactions to them, and the ways individuals who use hearts are understood. This includes how heart icons magnify and change individuals' feelings and encourage responses. Cultural criteria and many individuals' investments in pleasing others mean that such phrases as "I am in love with you" and "I lost my heart to you" are mandates for the addressed subjects to feel the same way and to reply, preferably with a version of the same impassioned phrase.

My review of people's expressions about their love for social networks, products, and members suggests how heart and love narratives render gender and sexual norms and queer attachments, including affection for brand communities. In the advertising industry, brands that generate impassioned feelings are sometimes described as "lovemarks."[14] Such companies as Etsy, Facebook, and Twitter work to attach people to their brands as a means of configuring individuals and sites. Due to the impassioned connotations of heart icons, people experience them as intense expressions and respond with similar intensities, including love and revulsion. Heart icons can also be part of the banal everydayness of online expressions and touchscreen engagements, which is distinct from intense and specific digital passions. Nevertheless, individuals in such social networking sites as Twitter, as I suggest later in this chapter, expressed extreme negative feelings when the site replaced the star with the heart button as a method of favoriting tweets. Twitter participants' concerns about having their identities tainted by hearting evoke the leaky interface that I discuss in chapter 2. I argue that such emotional and coded reactions demonstrate how individuals negotiate the feelings associated with online sites as a means of maintaining their identities. I also consider how sentiments reemerge as part of participants' refusals of such content. Thus, my analysis suggests ways of studying the connections between other online practices and feelings.

Configuring Feelings

Barthes offers a theory of intense embodied experiences, which can be related to the functions of heart icons and repetitive indications of loving people. He employs the concepts of punctum and studium to explain how photographs influence viewers, and he associates studium with the general societal aspects of images and surfaces that do not move him. Studium occurs when spectators only "take a kind of general interest" in images and engage through the "rational intermediary of an ethical and political culture."[15] Studium conveys a "body of information." Punctum "will break (or punctuate)" the cohesive photographic surface and the uniform cultural meanings of photography. According to Barthes, these temporary and unscripted ruptures emotionally influence the viewer and destabilize bodily coherence. His conception of photographic punctum, which has been identified as a form of affect and feeling, offers methods of considering how people respond to heart icons and how these symbols are related to and distinguished from the associated texts. Barthes's theoretical notion of punctum foregrounds the deeply embodied, unshareable, and queer experiences that individuals have with representations. His related arguments evoke many of the features of online heart icons and representations, including the ways they make the viewer passionately feel, pierce the text and viewer's experiences, and link the individual to representations, while being distinct from the more stable emotive aspects of these sites.

Heart emoticons, emoji, and related representations render a sensitive surface and interface. Such representations punctuate sites, messages, and surfaces with feelings. These effects are related to the aspects of sensitive photography and viewing that Barthes describes. In Elspeth H. Brown and Thy Phu's study of photography, they argue that Barthes conveys the complexities of feeling photography.[16] According to the photography theorist Shawn Michelle Smith, Barthes encourages individuals to have an "affective response" and to employ an "affective mode of approaching the photograph."[17] Barthes establishes a theory of feeling viewership by identifying punctum as "that accident which pricks," "bruises," and is "poignant" to the individual.[18] Punctum (and the heart) "rises from the scene, shoots out of it like an arrow, and pierces" the viewer. This "wound, this prick, this mark made by a pointed instrument" punctuates the photograph so that it is "sometimes even speckled with these sensitive points."[19] Barthes's evocations of the "prick" and excessive emotions cause scholars like Brown and Phu to argue that the punctum is queer and enacts a version of Barthes's gay desires, which are more overtly described in his posthumously published writing.[20] From these frameworks, I offer a critical model that can be used to consider the sensitive, feminine, and queer aspects of online hearts, including the ways normative subjects avoid the connotations of hearts. Hearts and their associated arrows, which may be implied even though arrows are ordinarily not represented along with online hearts, are imagined piercing and emotionally influencing the bodies and positions of viewers. Digital arrow-pointers and their linked hand-pointers also establish individuals' relations to documents and settings.

Barthes's notions of punctum and studium are further connected to the affective aspects of the digital by his indication that studium is "of the order of *liking*, not of *loving*."[21] Barthes's differentiation between liking and loving is related to distinctions between interface like buttons and the use of the heart to convey more impassioned feelings. Certainly, icons function at the level of studium, which Barthes suggests works to "inform, to represent, to surprise, to cause to signify, to provoke desire."[22] However, heart icons also create brief flashes and sensitive feelings in viewers, including excitement, anger, and embarrassment. The distinctions between Twitter's initial identification of the star as favorite and its later offering of the heart as like and love are, as I suggest in more detail later in this chapter, aspects of the site that generate a great deal of alienation and other emotions.

Hearts are also related to studium and more general societal formations because of the ways they configure individuals' feelings and specify the people who are expected to participate. Hearts are thus employed as part of what Woolgar describes as "configuring the user." He outlines how writers employ varied techniques to entice individuals and to "*control* the relationship between reader and text."[23] He relates such practices of configuring individuals to advertising and sealed packaging that guarantee that no one has tampered with products and that act as a form of consent to licensing agreements. Through such formulations, companies and designers represent users, instruct individuals about their needs and desires, and incorporate notions of users into products. These processes of configuration are productive for companies. For example, Etsy configures members as brides and lovers of weddings because of the consumer-oriented features of contemporary weddings and the cultural association of brides with wedding labor. These wedding figurations successfully mesh with Etsy's interests in having participants work for and love the site and the associated goods. Weddings also provide familiar and heartfelt notions that efface the economic, structural, and technological aspects of the site.

The human geographer Nigel Thrift extends Woolgar's thesis about configuring the user and asserts that electronic toys introduce children to interfaces, technologies, and the associated belief systems.[24] Electronic toys thus articulate expectations about children's adulthood and future engagements with sites and devices. Digital products and online sites configure participants through log-in and personalized messages, terms of service agreements, rule systems, instructional videos and texts, hearts and other mood icons, and representations of how the site is supposed to function. Touchscreen-based and other digital devices and online sites configure individuals' identities and communities' characteristics. These processes of configuration also delimit participants' actions and beliefs. For example, heart icons and related narratives about love tend to configure people's intimate relationships with sites and technologies as emotive rather than skill- or knowledge-based. Pulsing and bursting hearts are intended to convey, or even to produce, intense and escalating feelings in individuals.

Scholars have questioned the inflexibility of Woolgar's theory of configuration and suggested that the structuring of participants is a more flexible process that can be refused. Anne Sofie Lægran and James Stewart, who study new media, describe Woolgar's theory as a mechanistic assessment that privileges designers' perspectives over that of users.[25] Yet as I suggest in chapter 1, iPhone and other mobile phone buyers often support the positions of designers and manufacturers and enforce gender norms. Thus, participants' interests should be integrated into theories of configuring the user and gender scripts rather than remaining isolated areas of investigation. The research on gender scripts by such scholars as Madeleine Akrich, Majken Kirkegaard Rasmussen and Marianne Graves Petersen, and Nelly Oudshoorn, Els Rommes, and Marcelle Stienstra indicate how designers, companies, and individuals associate technologies with particular kinds of people.[26] Products convey expectations about who will use technologies, what kinds of bodies and interests will be recognized, and how devices will function. The associated gender scripts tend to be deeply normative and are the consequence of the meshing of technology producers and designers' worldviews with those of adopters and consumers. These scripts ordinarily correlate advanced technologies and technological skills with men. Practices and representations that are deemed to be frivolous, including shopping, chatting, and hearting, are associated with women. As these gender frameworks and the associated responses suggest, gender and other identity scripts are imbued with and generate feelings, including people's pleasure in identifying with technologies that are designed for them and anger and border patrolling when these privileges are interrogated. In this chapter as well as throughout this book, I demonstrate how online participants collaboratively and emotionally extend and dispute gender conventions and other technology scripts.

People often distinguish online configurations of sentiments from the feelings that are associated with ideal film viewers, who are rendered as masculine and male, and who are encouraged to take more distanced and critical stances in theaters. Mary Ann Doane considers and counters these constructions of ideal male viewers by pointing to how women are structured as being too close to their own representations.[27] Women have been historically associated with "weepies" and more recently with "chick flicks," which do not have the elevated connotations of mainstream genres. Younger and very engaged viewers are also seen as too impassioned and close to the screen when watching action and horror films. Cultural conceptions of distanced and elevated cinema viewers may be extended into other media environments via increases in the size of home screens and may be depleted by binge media viewing attachments and habits. Feminist film theorists highlight the ways the ideal male viewer is associated with the position of the apparatus and the empowered gaze of white heterosexual male protagonists, which render women as what Laura Mulvey describes as "to-be-looked-at-ness."[28] Hearts, as I suggest in this chapter, shift such frameworks and are more likely to emphasize feminine to-be-touched-ness. Companies' and online sites' employment of hearts should encourage scholarly acknowledgments of such practices and their influences in Internet, screen, and media engagements. For instance, my research on the functions of love and hearting would expand considerations of fannish interests and the ways weepies and chick flicks are gender coded as feminine and emotional.

Defining the Heart and Loving Feelings

The history of the heart is also a chronicle of touching and being touched. Researchers have proposed varied ideas about how heart-shaped icons were developed and associated with love and related sentiments.[29] The heart shape was employed as a means of representing foliage in examples from antiquity. During this period, people's use of the silphium plant and its heart-shaped seedpods and fruit as aphrodisiacs and contraceptives may have been factors in connecting this shape to love.[30] The heart shape is also widely related to the human heart even though they are visually and functionally different. As P. J. Vinken's book about the heart notes, the shape was employed in northern Italy in the early part of the fourteenth century and was informed by an error that Aristotle made in an anatomical text.[31] While anatomists corrected this mistake in the sixteenth century, the scalloped heart was already widely established. Starting in about 1480, French playing card designs included hearts.[32] Yet the Catholic Church, as Keelin McDonald reports, claims that the modern heart shape and its association with love and devotion appeared in the seventeenth century, when Saint Margaret Mary Alacoque had a vision of the Sacred Heart of Jesus, with a red heart surrounded by thorns.[33] Twentieth-century versions of the heart include the "I ♥ NY" campaign that promoted state tourism and was featured on T-shirts, mugs, and other goods.[34] In the New York tourism campaign, the heart is meant to convey the term "love," but it is understood as "heart" in other usages.

The consumer research of John F. Sherry and Mary Ann McGrath chronicles how marketing and shopping are connected to love. People render shopping as an amorous pursuit when they indicate that they "love" to shop and "fall in love" with selected goods, companies, and stores.[35] According to the literature on brand love, people are connected to brands through such experiential feelings as trust, passion, and commitment.[36] Rather than seeing these feelings as distinct from people's love for animate things, brand love makes the company and goods seem to be alive and highlights the ways brands are supported by workers, sellers, buyers, and animate attachments. Technology companies, as I suggest in the introduction, often render their products and people's experiences with devices as animate. People's embrasure of touchscreens and related forms of brand love make devices into corporeal skins and a corollary for individuals. In their research on brand love, Barbara A. Carroll and Aaron C. Ahuvia indicate that feelings for brands "may not be perfectly analogous to the feelings one has for other people" but they can be "considerably more intense than simple liking."[37] This too may shift between forms of Barthes's studium and punctum. Brand love, as Carroll and Ahuvia argue, "includes passion for the brand, attachment to the brand, positive evaluation of the brand, positive emotions in response to the brand, and declarations of love for the brand."[38] For instance, companies like Etsy reference love in slogans and other design elements.

Brand love is designed to intensify people's feelings for companies and products. The associated affective amplifications are often aspects of online settings and engagements. According to Susanna Paasonen, Ken Hillis, and Michael Petit, the term "affect" emphasizes intensity and describes a "quality of excess," or "more than."[39] They associate intensity with people experiencing sites and avatars as material and being delighted when finding desired prices online. Intensities might also be correlated with different-colored heart icons and the emotions that they convey, including varied kinds of passionate and fraternal feelings. For instance, Emojipedia identifies the solid red heart, which was the "Heavy Black Heart" before the development of color emoji, as a "classic love heart emoji, used for expressions of love."[40] The red heart's conveyance of a version of what Paasonen, Hillis, and Petit identify as excess should be related to the repetition of the term "love" in the definition.

In a variety of online sites, heart and love are emphasized, but so is hate. These frameworks for loathing may be intensified on sites that encourage anti-LGBTQIA+ sentiments, anti-Semitism, Islamophobia, misogyny, racism, and related forms of hate. For instance, the Stormfront site, which identifies its constituency as "White Nationalists" and the "new, embattled

White minority!" includes angry emotional icons that can be added to posts.[41] The site was created by Don Black, a Grand Wizard of the Ku Klux Klan and white supremacist, in 1995 and was the first major online hate site. It offers the ability to add "smilies," or "small graphical images that can be used to convey an emotion or feeling," to posts.[42] While there is a "kiss" smilie, there is no code for "heart" or "love." A representation of anger and varied expressions of bad feelings are available. Individuals can employ smilies to convey "Mad," "Banging head against wall," "smash" or "frustration," and four forms of "bang," where the smilie angrily shoots a gun at a target. Stormfront's smilies reflect and encourage a belief system and site that are focused on hating and harming dismissed and disenfranchised people.

Barthes's notion of punctum provides a useful means of intervening in Stormfront's violent smilies and the feelings produced by heart icons because it also "refers to the notion of punctuation."[43] Emoticons, emoji, and related symbols are often used to frame and conclude sentences. They also punctuate texts with feelings.[44] Physical expressions are limited in many online engagements. In written texts, including online communication, nuanced emotions and tone of voice can be conveyed through such punctuation as exclamation and question marks. Some reports have argued that using periods in emails and text messaging is too formal and lends an ominous valence to communication.[45] Employing all uppercase letters, as I have previously noted, is also scorned by many online participants as shouting. In Ariadna Matamoros-Fernández's Facebook research, she notes how the ability to intensify and amplify feelings through visual icons and other digital affordances has supported online hate initiatives.[46] The distinct aspects of digital communication have been addressed in varied proposals for conveying sentiments. For instance, the computer scientist Scott E. Fahlman recommended an early form of emoticons on a bulletin board in 1982 as a method of distinguishing between serious and joke posts.[47] The scholarly acknowledgment of these plans and an associated close reading of such things as repetitive punctuation, words, and emoticons can identify how feelings and other cultural belief systems are produced online.

People continue to develop specialized dictionaries as a means of producing and grappling with online terminology and representations. These texts have also structured the identity of online sites and digital cultures by articulating outsiders. The Jargon File, which was first produced in 1975, and the associated *Hacker's Dictionary*, which was published in 1983, include information about hacker culture and definitions for varied computer and Internet terms.[48] The current online version of the Hacker's Dictionary, which is an updated version of the print text, articulates the early and continuing functions of emoticons when indicating that they are "virtually required under certain circumstances in high-volume text-only communication forums."[49] The "lack of verbal and visual cues" in such settings can cause comments to be "badly misinterpreted (not always even by newbies)." With this proviso, the Hacker Dictionary also ranks people's technological and communication skills based on their time using online forums. Paul Andrews's *Seattle Times* article from 1994 provides negative views of newer participants and emoticons when informing the "newbie" that "Overuse of the smiley is a mark of loserhood!"[50] The "problem" of excessive emoticon use is associated with unknowing individuals when Andrews suggests, "Smileys have infected commercial networks as well, especially America Online." He identifies emoticons as nothing but noise and the "equivalent of crackling and popping on a cellular phone." Andrews's comments indicate how emoticons are represented as contaminating, especially when they are related to repetitive use. He conveys a technological bias in suggesting that the people who accessed the Internet through such systems as America Online did not understand digital settings and were not technologically skilled. In doing this, he provides a script about online expertise and class that is related to the gendered and raced scripts about dirty touchscreens that I consider in chapter 2.

Individuals' online employment of hearts are emphasized and sometimes dismissed in specialized dictionaries and encyclopedias. The online Emojipedia's search bar mentions the term "heart" and the smiling face with heart-eyes.[51] It thereby renders heart emoji as initial and exemplary cases. Emojipedia also encourages the reproduction of such symbols because a "Copy and paste this emoji" option is an aspect of each definition. Dictionary.com defines a variety of heart emoji and the specific feelings that they represent. For instance, it associates the black heart with "feeling angst-y," "misunderstood," and "emo" and having a "dark twisted soul, morbid sense of humor," or loving sad things.[52] According to Dictionary.com, heart icons articulate particular kinds of people and their feelings and behaviors, including emo-identified individuals who are engaged in subcultural music and fashion.

Dictionary.com indicates that "I ♥ u" can be understood as "'I love you' or 'I heart you'" and conveys the continued flexibility of such symbols.[53] This influence on the meaning of texts is based on where hearts are placed and how many are employed. For example, Dictionary.com identifies a heart at the end of a sentence as the "classic." It also describes people's tendencies to frame a "statement visually by putting a heart at both the beginning and end of it," to use a number of hearts for "emphasis," and to employ the heart as an "intensifier to express affection."[54] Heart icons decorate and elaborate upon members' online names and are employed in "social media handles in order to cultivate a certain type of online persona." For instance, autonomous sensory meridian response (ASMR) YouTube video producers use hearts and other symbols in their names and posts, as I suggest in chapter 4, to underscore their practices of loving and virtually touching viewers. Hearts intensify texts when people repeat the symbol and employ such icons as the beating heart. Online dictionaries also deserve critical analysis because they control and intensify the ways new media languages are understood.

Aly Trachtman posts to the student-oriented FlockU and suggests that "DOUBLE PINKS" hearts are employed "when you want to be girly, and flirty. Men who use this are probs gay, but weirder things have happened. This is definitely a 'go-to' color for women."[55] People's readings of hearts are based on varied contextual cues, but Trachtman specifies that hearts are associated with binary gender and sexuality positions. She indicates that this identity script is especially likely with pink hearts, which she correlates with femininity, women, and gay men. This correlation of emotions to gender positions and sexual identities is common. According to Steven L. Arxer's research on masculinity, emotional expressions are associated with femininities and nonnormative masculinities.[56] Emotions are believed to stigmatize men and normative masculinity because they are viewed as excessive and correlated with bodies that are seen as too much, including women, gay men, and people of color. Cultural norms insist that femininity should be connected to women, needs to be kept in check, should be enacted only in traditional ways, and is polluting when intensely performed or linked to unexpected bodies and things. Individuals use these structures to unfortunately suggest that men who engage in feminine things are gay and to indicate that femininity taints men and their masculinity.

Similar gender scripts are rendered on Urban Dictionary when craigums describes the <3 (or heart) icon as an "annoying way of putting a heart online mostly used by 15–17 year old girls who have nothing more to do than kiss boys that wear the same pants as them."[57] This poster uses dismissive

terms and utterances to suggest that teen women do not do anything important. The example for these women's speech is, "*OMGOMGOMGOMGOMGO
MOMGO MOGM i <3 his pants.*" craigums escalates this misogynistic narrative by describing the related phrase "lyke omg" as a "code used by 14–16
year old girls that has yet to be decoded" and suggests that women's interactions are illegible.[58] While repetitive icons, acronyms, words, phrases, and
punctuation are used for emphasis online, people's negative comments
and the correlation of these practices to femininity are some of the ways
women are deprecated. The associated indication that young women are
invested in insignificant behavior renders stereotyped distinctions between
women's frivolous technology engagements and men's more productive
activities. These gendered hierarchies are also established in misogynistic
narratives about women's purportedly inadequate mobile phone use, as I
suggest in chapters 1 and 2.

Scholarly studies of gender scripts can be employed as methods of critiquing such texts. In this chapter, I further develop gender script literature
to address identity scripts and suggest some of the ways these inquiries can
be employed in considering the forms of communication that are associated with specific subjects, the ways gender scripts are intermeshed with
race and sexuality scripts (such as linking women to white weddings or
nonnormative men who wear the same pants), and how repetition functions. The related analysis might also address the ways the utility and
multiple functions of symbols and texts are condensed and rigidified. For
instance, the digital rhetoric scholar Elizabeth Losh's study of hashtags is a
reminder that the # symbol, which some people now only associate with
social media tagging, predates computing.[59] It supports nonhuman reading,
as well as individuals' social media tagging, activism, and articulations of
collective beliefs and sentiments. This includes enactments of and attempts
to control feminist hashtag activism, which I address in the afterword. Yet
the politics of hashtags and hearts, and the tendency to use these symbols
multiple times, are displaced by Urban Dictionary posters' rendering of
them as feminine, annoying, and noncommunicative.[60]

The value of feminine communication is further dismissed on Urban
Dictionary when Birdypwnsjoo argues, "<3 has been so overused it doesn't
even mean what it looks like anymore."[61] Birdypwnsjoo thus identifies the
heart as ambiguous, worn out, and deceptive. Ulrecht indicates, "Apparently it means love or heart. It's all over Twitter at the end of girl's posts."[62]

The use of the term "apparently" works to make it exceptionally obvious and distances Ulrecht from what are represented as excessive and overused bouts of feelings. The symbols themselves are "all over," and thus understood as too much and contaminating. While Barthes indicates that the surprise of photographic punctum is supported by its processes of coming into being and evanescence, the <3 is sometimes figured as overexposed. As Birdypwnsjoo and Ulrecht propose, the employment of multiple hearts is often culturally dismissed as too insistent, too visible, and too repetitive, and thus as too feminine and queer.

Femininity and feminine excess have been understood by some second- and third-wave feminists as attempts to conform to male interests and normative heterosexuality, but the posts on Urban Dictionary highlight the negative connotations of feminine excess.[63] Adele Patrick's research on feminine excess identifies how the "elaborate fashioning of femininity, in hair styling, clothes, and cosmetic acts has been read variously as 'excessive,' counter-revolutionary, radical, inappropriate, and 'queer.'"[64] As she indicates, there are political utilities to enacting such positions when they distort expectations. This includes women's refusal to establish themselves as appropriately classed and tasteful. Patrick suggests that women's enactments of excess femininity can undermine cultural expectations of stable gender roles and thus function as a form of Sara Ahmed's feminist killjoy.[65] However, craigums tries to enforce stable binary gender and sexuality positions when associating <3 with what are supposed to be women's disproportionate and misconceived desires because of their interest in men who wear the same pants as them.

Men who wear women's clothing are often culturally demarcated as engaging in inappropriate gender practices and being queer and gay. Sarah Banet-Weiser and Kate M. Miltner's research foregrounds instances where heterosexual masculinity is challenged by femininity.[66] Given the association of the heart with femininity, the linking of the <3 with a group of men who wear pants that are also adopted by women is designed to undermine the associated men's positions as normatively masculine. For example, Andre defines the <3 as "an emoticon for a really gay heart" and proposes that users and receivers of the associated messages are tainted by such content.[67] Since the heart has been employed as an intensifier, definitions on Urban Dictionary suggest that it results in the production of insistently and "really" gay individuals. These formulations point to how gay identities and hearts are correlated in their cultural roles as magnifiers of characteristics and feelings. In these settings, gay sexualities are culturally stereotyped as some combination of being too feminine, too masculine, and inappropriately gendered. Thus, these gender scripts that associate women with hearts are reliant on sexuality scripts. These practices can also be considered, and their problems pinpointed and punctured, by employing Barthes's conception of punctum and other affective queer theory frameworks, which I outline later in this book. Barthes's notion of the pricked and torn subject compromises investments in impenetrable and coherent bodies and the associated conceptions of normative masculinity.

Consumer Love on Etsy

Ecommerce sites that are not actively directed at heterosexual men tend to emphasize femininity. Sites that are focused on women and femininity are more likely to employ hearts to underscore that they love people and that these emotions replace the impersonal and massified aspects of online shopping. Etsy and related online sites associate the heart and love with handmade objects, items crafted or sold by passionate individuals, things that reference personal experiences, and items that are gifted by intimates. In *Buy It Now: Lessons from eBay* and *Producing Women: The Internet, Traditional Femininity, Queerness, and Creativity*, I note how companies, sellers, and buyers work together to associate the online marketing of such goods with personal connections between people.[68] These companies and individuals indicate that handmade items connect people through the shared touch of objects, are imbued with loving feelings, extend familial forms of love, and are pleasurable to create. For instance, craft producers who identify as mothers often indicate that they are selling products that are similar to the ones that they lovingly made for their children and sharing their related feelings. Christoph Fuchs, Martin Schreier, and Stijn M. J. van Osselaer's research on handmade objects indicates that these "products may often be more attractive, at least in part, because they are perceived as being made with artisanal love and even as symbolically containing love."[69] They "define love as the producer's warmhearted passion for a product or its production process that, as a result, can be perceived as symbolically embedded in the product." Thus, people understand handmade objects as representing, filled with, and extending circuits of feelings.

Etsy foregrounds these notions of consumer love in its email sign-up for wedding information, which has been available in the same form for many years (figure 3.1).[70] The sign-up site features a photograph of a white heterosexual couple holding a banner. The banner, which loops under and seems to conceptually support the sign-up form, has pendants that spell out the word "LOVE" and hearts that resemble Etsy's favorite button. The form promises to provide "picks for brides, grooms and everyone else who loves weddings" and scripts people's interests as being wedding-oriented. These representations connect Etsy to the processes of heterosexual coupling, loving individuals and goods, and what is presumed to be a more general love of weddings. This focus on weddings, which as I suggest in *Buy It Now* is also employed by eBay, allows ecommerce sites to reform the feminine, queer, and excessive connotations of shopping and associate the site and company with sanctified, heteronormative marriages and familial attachments.[71] Etsy's scripting of the wedding as a site where white heterosexual and community attachments are produced is counter to its promise to acknowledge and connect everyone through their love of weddings.

Etsy's digital banner, which is on the initial part of its site, adopts a related narrative about emotional connections when directing individuals

Figure 3.1
Screenshot from "Etsy – Email Sign Up."

to "Buy directly from someone who put their heart and soul into making something special."[72] Etsy employs this account about the heart to indicate that sellers personally put everything into objects and give everything to consumers, including their passion and care. The site uses a form of tactile address by directly engaging buyers and referencing how producers have emotionally touched items. This structure is designed to get individuals to invest in the brand. Buyers are supposed to return the site's and sellers' sentiments by purchasing items and supporting Etsy's construction of an ethics of care. Through Etsy's text about the heart and soul of special items, the company associates the process of buying massified objects or purchasing through other consumer processes with not caring. Another version of this ardent relationship is constituted when Etsy tells readers to "Support independent creators. There's no Etsy warehouse – just millions of people selling the things they love." The site figures a version of Diana Adis Tahhan's touching at depth in which the company helps "you connect directly with makers" and share deep feelings with sellers and their products.[73] Etsy members are assured that they are in direct contact with makers and will receive the things that sellers love. Viewers are also configured as people who care about and make sellers' economic sustainability viable. Yet, as I suggest in *Producing Women*, online sales of personally produced and handmade items do not always result in viable jobs.[74]

The people and companies marketing handcrafted goods, including Etsy, suggest that handmade items connect people in emotional and tactile networks that are not based on physical contact. For instance, Etsy entices individuals around Father's Day to send "Dad love from any distance."[75] In a related manner, Tahhan argues that "heart (as feeling) really signifies a presence of touching at depth, touch that is not locatable at all."[76] These feelings, which include such notions as hearfelt and heartwarming, incorporate a "tug or warmth—a tangible feeling of connection." Cultural conceptions of heart include emotional associations, exchanges of feelings, and embodied responses. These sensations are then supposed to be felt "inside" the body. People insistently identify online associations, including ecommerce exchanges, as intermeshing heart, love, and touch. So too do conceptual and coded links mesh bodies, avatars, companies, devices, software, sites, and texts together. According to Tahhan, touching at depth "has intimate manifestations (is not only physical or from the 'body') and finds meaning through an embodied felt relation and deep sense of connection."[77] Throughout this chapter and book, I elaborate upon Tahhan's consideration of the ways touch is emotionally experienced by foregrounding the stakeholders who produce and experience such touching at depth despite (or because of) distance. This also allows me to indicate the ways touching at depth can be employed to theorize the representations and sensations associated with digital media.

Etsy's About page emphasizes the importance of touching and crafting at depth, including the ways these experiences contribute to its site ethos, by representing a series of animated and engaged hands. Two hands hold knitting needles and ply the yarn in unison (figure 3.2). A lighter and darker hand touch palms. A light-skinned hand also swipes a touchscreen and changes the associated image. Such hands support Etsy's claim that it collaborates with buyers and sellers to "Keep Commerce Human" and individuals connected at a distance.[78] Philosophers and scientists have historically identified hands, as I suggest in the introduction, as the physiognomic features that distinguish and elevate humans over other animals. Etsy continues this association of hands with humans, while promising

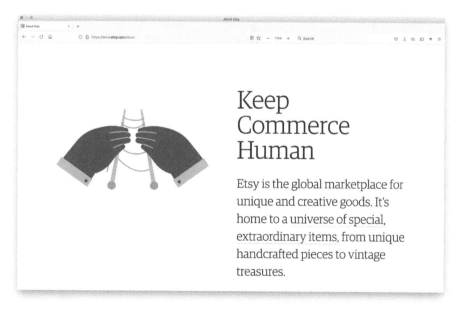

Figure 3.2
Screenshot from "About Etsy."

participants physical and emotive connections through the interface. Etsy also uses these hands to conceptually connect viewers to "unique hand-crafted pieces" in a setting where the texture and details of the handmade are difficult to determine. Manipulable hand-pointers and other operable and static representations of participants configure expected bodies and engagements. Through Etsy's representations of hands and contact, the company elides mediation and emphasizes the physical, tactile, and emotional aspects of its site and products.

References to hands and hearts make inaccessible goods into tangible things and sentimental connections. They suggest that buyers, sellers, and the company constitute a loving community where skin and bodies are in contact. Etsy's images of hands with tools indicate that the site facilitates people's processes of making things. Etsy identifies the site as a "vibrant community of real people connecting over special goods. The platform empowers sellers to do what they love and helps buyers find what they love." These narratives about hands, love, and hearts undergird the site and the ways it is supposed to make individuals feel. In a similar manner to other social selling platforms and social networking sites, individuals are encouraged to identify with and contribute to the brand and interface and understand these arrangements as a community.

Lisa Whitmer encouraged such connections in her role as manager of Etsy's seller education team. She notes, "One way to turn browsers into buyers is to hook them" with a video about "what makes your Etsy shop noteworthy."[79] Her mention of "browsers" combines the viewing technologies that allow individuals to access sites with people's processes of engaging. She also articulates physical interconnections in the form of a "hook." Sellers are encouraged to share by answering the question, "Why do you love what you do?" In reply to her prompt, sellers respond with related sentiments. For example, Louisa writes, "love that this is now an option on etsy! Inspired to create a video soon!"[80] This seller employs the term "love" to convey extreme happiness. Pleasure and the movement of loving feelings through varied circuits are also evoked when Adriana notes, "love this!"[81] The video option will allow her "Etsian loves" to see her shop and view her explanation of her process. She also expresses her "love" of "Etsy!" and how she would "love" to "connect" by seeing buyers' videos. Her continued employment of the term "love," as well as exclamation marks and related sentiments, in a similar manner to my considerations of other forms of repetition, are intended to convey her intense feelings and notion that the site provides intimate and exciting connections.

Heart Buttons and Hearting on Etsy

Individuals may be hooked and connected by buttons that evoke touching and feeling. Heart buttons are increasingly available on ecommerce sites, including sites that are as distinct as eBay, Etsy, Modcloth, and Neiman Marcus. Buttons are different than Hypertext Markup Language (HTML) links, which indicate that individuals can open a web page or resource. Buttons suggest that individuals can modify data or add preprogrammed indications. Buttons reference the actions of hands and fingers on surfaces by suggesting that there is something to touch and push. They thereby configure the expected body of participants and the position of individuals' hands and fingers. Heart buttons may thus be particularly productive for Etsy because they stand in for certain kinds of material work. Yet button pushing has also been figured as mechanical and excessive in its requirements for force and repetition. Rachel Plotnick provides a historical chronicle of people's concerns about material buttons being replaced by images and screens. According to Plotnick, individuals have conveyed apprehensions about buttons being replaced by smooth glass and "force and feedback" barely mattering.[82] While she relates such screen-conveyed buttons to "flatness and touch without feeling," I suggest that heart and other buttons and associated options render tactile and emotional sentiments. Given buttons' recurring and expressive functions, they are related to individuals' repetitive use of texts for emotional impact and thereby to a kind of familiar excess.

Varied kinds of buttons, including plant buds, pimples, belly buttons, and genitals, are conceptualized as tender parts. On-screen buttons are also pleasurable and sore points and thereby evoke Barthes's conception of punctum. People are more likely to experience such soreness when they refuse heart buttons' conveyance of wholehearted feelings and the repetitive spotting of the interface with heartfelt sentiments. Punctum and hearts, including Valentine's Day stereotypes, seem to shoot out at viewers and produce holes. Digital buttons call to and provide a point for the fingers of users, or their hand-pointers, and position bodies in relation to devices and screens. This is enacted when individuals put their fingers on touchscreens. Buttons address the "you" who will engage with the system and thus produce a direct and tactile address. While Ahmed figures the raised arm as a position of feminist resistance, buttons provide specific directions and schematics for hands and fingers. A theory of buttons, which I outline in this chapter, should be incorporated into studies of online interfaces and mobile touchscreen devices because buttons promise interactivity while scripting the actions and bodies of individuals.

Heather Burkman's Etsy blog post provides a context for heart button pushing that connects tender feelings to collecting and buying. She informs readers, "By clicking the heart, the item will be added to your default Favorites list, 'Items I Love,' for fast collecting."[83] Burkman and Etsy directly address individuals and deploy the button structure to articulate people's actions and feelings. While individuals may add items to a list for varied reasons, Etsy's terminology and framework render items as coveted goods that generate affectionate feelings. Buttons directly address individuals as people who should select the highlighted options and, in the case of heart buttons, understand their interaction through enthusiastic, happy, heartfelt, and loving sentiments. These online buttons are often surrounded by and conceptually associated with emoticon and emoji hearts and narratives about hearting and loving. Yet notions about the people who use buttons are also proscribed by interface articulations of white hands and other cultural expectations about users.

Sellers, as I have started to suggest, support Etsy's valuation of the heart, hearting, and loving feelings. For instance, dfalv38 conveys excitement when getting her "100th Etsy heart today!!! Yippee!!"[84] This seller envisions turning "hearts into sales and lifelong customers" and underscores how hearts contribute to social selling. She also indicates that people "exchange hearts" in forums. The tactical exchange of hearts for profit and the boost of the likelihood that people will notice a seller may invoke the black and cold heart that is often associated with lack of feeling. Yet dfalv38 employs and amplifies the heart icon when using a series of black hearts as a text divider. After this emotive break, dfalv38 writes, "Everyone on Etsy wants to know how to create these little black heart symbols to add to their descriptive copy" and "make neat bullet points." dfalv38 provides instructions on varied ways to produce hearts, including cutting and pasting. She thereby makes her post into a template for readers' practices and repetitions. dfalv38 suggests that using hearts is a creative enterprise that supplants the Etsy interface and functions as emotive bullet points, punctum, and economic stimulus.

EtsyGadget's Hearts Counter facilitates such practices. It enables sellers to combine the heart button with a plug-in that "shows how many times have you favorited/unfavorited (hearted/unhearted) items and shops during an hour."[85] Since "many sellers participate" in "'Favathons,' the main principle of which is heart interchange," the plug-in allows individuals to "prevent missed hearts and so, avoid being expulsed from teams." Hearts are thus a means of exchange that connote favorite things and can result in interpersonal problems when not delivered. People respond to the Hearts Counter with related narratives about love. These textual forms of call-and-response suggest how related ideas are produced. Rose Barbola replies, "Wonderful tool! Loving it!"[86] Pamela Quinn from Vintagequinngifts writes, "love the heart counter."[87] Their use of the term "love" is meant to convey appreciation of the application and its delivery of information. They indicate how intensities are expected, produced, and circulated as part of heart exchanges.

Sellers also associate Etsy, members, and items with feelings in the commentaries that are part of their stores and listings. Sellers thereby follow the scripts that I have analyzed, and that Etsy established. For instance, Val Hebert identifies as a "multimedia artist who is currently sidetracked by a love affair with fiber."[88] She relates her passion for crafting to human interactions and pleasurably excessive relationships with objects. Hebert identifies herself as a "painter at heart" and renders the heart as central. She also "loves bringing her sensibilities together on the canvas of repurposed wool, which she needle felts" to her "heart's content." Hebert thus distinguishes her work as consolidating her feelings and fulfilling desires. Her practices and products are also supposed to meet buyers' yearnings, and it is "100% guaranteed that you will love what you buy!" In writing this, Hebert promises to fulfill Etsy's indication that individuals will "Find things you'll love."[89] Hebert offers a "thank you from the very bottom" of her "heart for supporting H a n d m a d e."[90] Her expressed appreciation advances the site's constitution of deep feelings and touching at depth.

Hebert uses a large font to offer consumers "HeARTfelt goods made of reclaimed textiles, embellished with needle felted love, one handmade piece at a time."[91] She references multiple forms of touching at depth, which are associated with and supposed to be conveyed through the feeling of materials, including the felt fabric. An analysis of such terms is supported by Barbara Johnson's identification of the importance of addressing ambiguous words.[92] Felt (fabric) provides an ideal term for evoking emotional feelings that combine with and are magnified by tactile things. The fabric is physically and emotionally felt because of its texture and because it is pierced with the producer's handwork and interests. Needle felting is the practice of consistently pushing a sharp object into a fabric until the fibers knot and weave together and become felt fabric. Such processes render a constructed version of punctum because the producer pierces material over and over again with a needle in a manner that creates felt and facilitates consumer desires. Hebert's chronicle of this piercing is designed to convince consumers of the labors of love that she incorporates into her products, and for her to indicate that she and her goods pierce buyers' hearts in a further form of punctum. Her description and process also indicate the ways entanglements, including the meshing of fibers, produce new structures and feelings.

Hebert's linking of feelings to enmeshed fibers and felt fabric evokes Didier Anzieu's identification of skin wrappings (and embodiment) as communicative.[93] It also conjures Steven Connor's and Tarja Laine's resonating membrane, which I mention in chapter 2.[94] Their proposals suggest the ways Hebert's multiple meanings and layers of felt are designed to make individuals collaboratively feel. Hebert and many other Etsy sellers use tactility as a means of connecting with other people. For instance, Elvira Para employs the tagline, "Creating useable, touchable art."[95] Rather than viewing her art at a distance, she wants people to "be able to touch it and use it in their daily lives." Para enjoys when people walk into her "show booth and immediately pick up a coaster, run their hands over a smooth satin tile." She emphasizes the ways textural objects attract individuals and make them want to feel things. In describing these connections, she encourages potential online buyers to imaginatively reach out, connect with her, and buy her ecommerce products. Thus, narratives about materiality are designed to reframe online shopping and make buyers feel as if they are reaching through the interface and touching art and other things. Such texts also shift women from existing as to-be-touched-ness to producing it, especially the rendered to-be-touched-ness of their crafted objects.

Sellers connect crafting to feeling by highlighting the ways the terms "art" and "heart" share a series of letters. Here too, the evocative aspects of words and the larger and thus more impassioned font are employed for their multiple connotations and associated sentiments. For instance, Hebert constitutes "HeARTfelt goods" that intermesh the heart, art, and feelings.[96] The Heartmaker and associated Heartistics shop produce these connections by crafting heart-shaped items and identifying the crafter as a "Heartmaker."[97] Beverly Thomas Jenkins expresses her "LOVE LOVE LOVE" of "DOING MOSAICS!!"[98] All of her "mosaics have a heart on them," which is identified as her "heart" and which she inserts into each item. The mosaics have a heart visually incorporated into the design that stands in for Jenkins's body, emotional feelings, and ability to touch shoppers at a depth by shipping a version of her heart to buyers. These conveyances of love, care, and appreciation are concluded with a "Thank you...love, Beverly xoxo." The intensity of these sellers' feelings and the overall resonance of sites are amplified through multiple uses of the terms "heart" and "love" and intensifying punctuation. Connor's and Laine's identification of skin as a resonating or beating membrane also suggests the ways figurations of hearts are designed to connect sellers and buyers.

The Heartmaker advertises the "perfect heart for the hearts you hold" and emotional and material exchanges.[99] This suggests that the seller, buyers, and people who are gifted these items are connected in a tissue of love and care. The Heartmaker's craft "fills" the crafter's "very heart" and moves from that "heart to yours."[100] The seller's bodily sensations are supposed to be transformed through the processes of creating things, and these embodied adjustments touch buyers and change them. This experience of being touched at a distance, and without physical contact between bodies, is echoed by and resonates with buyers. For example, clburon1234 describes wearing the pendant over the individual's "own heart everyday."[101] This buyer has and feels The Heartmaker's heart against the buyer's own heart. These texts indicate how consumers follow the scripts and experience the sentiments of sellers. The narratives also demonstrate how Etsy and its sellers produce sticky affective intensities that keep individuals engaged with the company and producers and link individuals in circuits of feelings.

Facebook Reactions

Etsy claims that its site emotionally connects sellers, buyers, and handcrafted objects in exchanges of loving feelings. Facebook is also associated with positive feelings. Caitlin Dewey reports on "Facebook's cultural and algorithmic preferences for everything positive and upbeat."[102] In a related manner, Lin Qiu, Han Lin, Angela K. Leung, and William Tov's Facebook research indicates that individuals are more likely to use the platform to convey positive emotions and well-being.[103] These good feelings are often conveyed through the love and care reactions, which include hearts. Yet scholarly research specifies that Facebook's interface goes beyond supporting friendships and articulates the terms and conceptions of relationships. According to Corina Sas, Alan Dix, Jennefer Hart, and Ronghui Su, "at their heart, people's most memorable experiences with Facebook are all about positive emotions."[104] These researchers emphasize not only how Facebook is correlated with upbeat feelings but also the ways the researchers identify such recollections as having essence and heart.

Facebook promotes positive feelings through its interface design, narratives about the site and its members, and reports about site changes. For instance, Facebook asserts that it added a series of reactions to the thumbs-up "like" option as a means of expanding upon the site's emotional features. Facebook's expansion of its like button in 2016 included options for "love," "haha," "wow," "sad," and "angry." In a similar manner to other sites, a graphic heart is used to represent love. The product designer Geoff Teehan indicates that the company wanted to "make the Like button more expressive."[105] Fidji Simo, who manages applications, elaborates on the association of the reaction buttons with various sentiments. The company launched reactions because people sought to "express how they felt."[106] Alexandru Voica amplified these feelings as part of his communications position, when he announced that Facebook was launching a new care reaction as a "way for people to share their support with one another during" the coronavirus pandemic.[107] This decision represents Facebook and its members as caring and foregrounds tactility because the icon is a yellow smiley face-like figure that is animated to tenderly hold and hug a heart, and by analogy also embrace the participants.

Facebook introduced the reaction buttons in response to concerns that the company had been grappling with since incorporating the like button in 2009. As Robinson Meyer reports, "Facebook has known that not every kind of post deserved a thumbs-up."[108] Mark Zuckerberg, the co-founder of Facebook and Meta Platforms, argues that people want the "ability to express empathy. Not every moment is a good moment."[109] This linking of reactions to bad moments is undermined by Facebook's tendency to render and promote positive feelings, including its mediation of the coronavirus pandemic with the figure that hugs a heart and is designed to send positive feelings and connect with recipients. This notion of care is part of Facebook's ongoing structuration and corporate identification. For instance, the company calls part of its research and development practices "Compassion Research Day." During one such research event, staff considered a "Sympathize" button, which may now be realized as the care button. Meyer argues that reaction buttons are similar to this figuration and offer the opportunity to express sympathy. The envisioned sympathize button and eventual development of reactions are designed to intensify the relationship between members and the brand by providing additional ways for people's feelings to emerge from the interface. Reactions enable participants to engage in the kinds of brand community practices, which Albert M. Muñiz and Thomas C. O'Guinn describe, and share in the sentiments and ethos of the site.[110]

Facebook Guidelines provide a similarly upbeat portrayal of reaction buttons while encouraging standardization. They note, "Reactions are an extension of the Like button to give people more ways to express themselves."[111] Yet Facebook informs participants that it is important to "use Reactions in the way they were originally intended" as a "quick and easy way to express how you feel." The company and site attempt to manage reactions as a means of encouraging people to understand the interface as authentic and as facilitating unmediated feelings. This is also the way YouTube scripts viewers, as I indicate in chapter 4. In fostering this framework, Facebook configures participants' practices even as it tries to hide how the site produces and mandates certain behaviors. Yet Facebook declares its interest in protecting the "integrity of the Facebook brand and product" rather than facilitating the widest array of members' feelings and expressions.

Facebook's control of members' feelings and identities includes banning drag queens, Indigenous people, and trans individuals for ostensibly not following Facebook's policy on "real names."[112] The #MyNameIs campaign, which is associated with the Electronic Frontier Foundation, asserts that Facebook's name policy is "culturally biased and technically flawed" and represents corporate indifference.[113] The policy endangers individuals who need to remain anonymous or pseudonymous because of domestic violence, stalking, repressive government, and political activism. People duplicitously use the reporting option as a means of silencing groups and individuals. Facebook has banned or otherwise suppressed individuals who do not meet gender and racial norms, even as it has been a welcoming site for online hate, the spread of violent content, and misinformation.

For instance, Brenton Harrison Tarrant is alleged to have used Facebook to livestream his white supremacist terrorist attack on New Zealand mosques.[114] White nationalists also used Facebook to organize the Unite the Right rally in Charlottesville, Virginia, in 2017 and more recent events.[115]

Ariadna Matamoros-Fernández's Facebook research indicates that people use the angry reaction as a means of producing and amplifying racism and hate. As she notes, far-right groups have "appropriated this affordance to spread anger towards specific targets."[116] Matamoros-Fernández argues that such platform affordances "facilitate the mobilisation of anger as a tool of power." Certainly, anger and associated threats of violence continue to be used as methods of silencing women, people of color, queer folk, and other oppressed groups. These feelings are amplified, and pleasure is correlated with inflicting harm, in Stormfront's association of smilies and things that bring smiles to people's faces with spraying individuals with bullets. Stormfront's use of smilies is consonant with its white supremacist message and should encourage an analysis of how smilies and emoticons map onto cruel, duplicitous, and hateful smiles as well as good feelings.

Facebook's banning of people's self-representations and enabling of hate through system affordances indicate the limits of its care. This suggests that its rollout of the care reaction may have more to do with downplaying news coverage about its escalating association with hate groups rather than addressing the coronavirus pandemic, including the disproportionate risk of death and serious illness experienced by people of color. In Andreas Chatzidakis, Jamie Hakim, Jo Littler, Catherine Rottenberg, and Lynne Segal's article about the ways care is referenced, enacted, and withheld during the coronavirus pandemic, the authors argue that corporations are attempting to increase their acceptability by self-representing as "socially responsible" at the same time as they contribute to oppression.[117] In the case of Facebook, the company markets and benefits from the crisis in care (and intolerance) by offering reactions, even though the site has not cared enough to moderate white and male supremacist posts, stop the circulation of misinformation, or intervene in the hate that is directed at other members.

Sammi Krug has managed the site's legitimacy and claim to caring in her role as Facebook product manager. She indicates that the site provides updates on "anything that matters most to you" and is the "central place to have conversations with the people you care about."[118] Krug portrays the addition of reactions as a process of "listening to people" and providing "more ways to easily and quickly express how something you see in News Feed makes you feel." Facebook consistently uses the term "feel," emphasizes members' sentiments, and indicates that the interface supports participants' moods. Krug and others identify reactions as ways of conveying more fine-grained feelings and engaging in more specific, quicker, and more abbreviated expressions. Reactions convey emotive responses to posts and other members and are designed to articulate the relationship between participants and the site. They also reshape members' practices of engaging with individuals and feelings about the interface. Dewey reports that the reaction options "have made it far more pleasant to respond to certain types of posts – particularly sad or outrage-inducing ones. They've also, like virtually all features on Facebook, made it easier for the network to amass data."[119] Dewey connects the ways Facebook enables people to feel good about bad feelings to company profits. In addition, Facebook intensified members' alienation and concern about such data collection when it admitted to changing the number of positive and negative posts seen by more than half a million individuals as part of a psychological study.[120]

A group of Facebook employees and reporters have emphasized the expressive aspects of the newer reaction options. The link between the like button and feelings is underconsidered and requires more analysis. Distinctions between the like and reaction buttons enact a version of Barthes's differentiation between general cultural experiences of liking and more embodied experiences of loving. In some of the reporting, the like option is understood as a less emotive response and one associated with standard behavior. This is conveyed by Connie Fredrickson's "Every time it's acceptable to use the new Facebook reactions" article about the expansion of the like button, which includes a photograph of a hand with an extended thumb and chipped blue nail polish (figure 3.3).[121] The blue nail polish evokes the blue of the Favicon, or shortcut icon, and the background color of the Facebook interface. The condition of the nail polish suggests that Facebook, and the referenced like button, are worn and in need of upgrading. This proposal that the like button is dated is supported by the subtitle of Fredrickson's article, which reads, "Like: The 'vanilla', safe yet a bit boring, of the Facebook reaction buttons." The dismissal of "vanilla" and "safe" (sex) likes renders reactions as kinky, a queer erotic, and a kind of

Like

Figure 3.3
Screenshot from "Every time it's acceptable to use the new Facebook reactions."

online barebacking. On Urban Dictionary, The Smart 1 similarly identifies a "vanilla like" as a "plain, lame Facebook reaction that normies consistently use."[122] By describing the like as a banal norm, The Smart 1 and others suggest that reactions are queer expressions that do not match normative identity positions. Of course, the rendering of hearts as deeply feminine tends to support binary categories.

Dewey elaborates on the emotional limitations of the like option when she argues, "Likers are indifferent" to "your post."[123] They acknowledge having seen it but "couldn't really be bothered to work up much of an emotional reaction. Likers are cool, distant, dispossessed." She suggests that many people's engagements, and presumably features of the interface before the redesign, are unresponsive and outdated. Her formulation also implies that liking is antithetical to touching at depth. It is notable that Dewey suggests that Facebook's moving hand does *not* evoke connections. However, virtual touch is amplified, as Brian Adam reports, with the care reaction, which shows "who you want to be close to despite the distance, it is sharing love during the quarantine."[124] Dewey also associates responsive processes with people's employment of the heart when you "'like,' but with feeling." This is unfortunately combined with stereotypes when she describes the love reaction as the "high-pitched 'omg, love it' of reaction emoji." She thus denigrates more expressive instances of conveying feelings by associating such reactions with the excessive, shrill, and feminine. Dewey's illustrative phrasing renders the love icon as disproportionate *and* uncommunicative. As I suggest earlier in this chapter, "omg" is deemed to convey immature and unclear feelings through speech. Urban Dictionary and other online sites specify that acronyms like omg are correlated with young women and devalued.

Other reporters indicate that there are significant limits to communicating with reactions and the availability of these options curtails other kinds of interaction. Natt Garun asks, "why bother commenting on a comment when you can just click the 'Heart' emoji?"[125] Garun suggests that contemporary subjects, or at least the feminine participants who are correlated with the heart button, have been made into automatons that mechanically respond. He writes, "What is communication even? Words? What are those?" In such instances, the heart is represented as magnifying the simplistic aspects of push buttons, but also imbuing the purportedly flat and affectless aspects of screen buttons, as narrated by Plotnick, with too much feeling.[126] Frank Bank conveys a similar position when noting that "more expressive communication" can be conveyed through words than through likes.[127] In a similar manner to Dewey's disaffected participants who "like" but do not care or engage, Garun and Bank articulate subjects who are prelinguistic and postlinguistic. Such narratives produce gender scripts about excessive and illiterate women participants. The women who use the heart reaction are depicted as thoughtlessly adopting empty signs and reducing already phatic communication to its most limited expressions. Yet N. Katherine Hayles emphasizes the intricate ways people communicate and write into and through online systems.[128] People's use of ambiguous concepts, multilayered meaning, atypical punctuation, emoticons, and emoji, as

I indicate throughout this book, indicate individuals' skills with and complexification of online communication.

People's multifaceted use of feminine hearts *and* individuals' contrary desires to contain such identities and expressions result in the persistent belittling of the heart, the sentiments that it evokes, and the participants who are correlated with love reactions and feelings. This may be because, as a group of journalists and other participants note, the heart is persistently employed. For instance, Jylian Russell suggests that 2016 was a "traumatic year" because of the election of Donald Trump as 45th president of the United States, but the "Love button" rather than another reaction "(fittingly) got the most love."[129] Karissa Bell reports that "'love' is far and away the most popular of all reactions."[130] These distinctions between dismissals and celebrations of hearting underscore how the heart represents and generates diverse feelings. In the next section, I consider people's expressed refusals of the heart on Twitter. Whether people happily adopt or angrily reject the heart and other site frameworks, impassioned expressions are productive for both sites and brands. People who respond with the love reaction, as Felicity Wild indicates, "can be assumed to be loyal customers and important brand ambassadors."[131] Negative evaluations can cause problems for a brand's community, but such assessments also act as reminders about the company and its products. The love button is valued and dismissed for its conveyance of members' identities, attachments, and amplified feelings. Its popularity and massification are part of its productivity and the reason some individuals suggest that it is too much, including its persistent appearance.

Twitter and the Broken Heart

The emojitracker tends to list the heavy black heart, which now appears on many sites in red, as one of the most popular Twitter emoji.[132] At the bottom of the emojitracker page, the centrality of the heart and the associated conveyance of feelings are supported by the indication that "emojitracker is brought to you with ♥" through the programming of mroth. The emojitracker's extensive grid emphasizes that hearts and other emoji are employed as part of everyday communication. However, a large number of journalists and other Twitter participants expressed extreme dissatisfaction when Twitter changed the star "favorite" button to the heart "like" button in 2015.[133] While many people disparaged Twitter's shift from the star to the heart, the change did more than make the site consonant with other social media platforms. The heart suggests that Twitter is a setting of touching emotions rather than preferences.

Twitter explained this interface change by tweeting, "You can say a lot with a heart. Introducing a new way to show how you feel on Twitter."[134] A Graphics Interchange Format (GIF) animation accompanies the tweet and depicts a throbbing red heart that equals "yes!!" "congrats," "LOL," "adorbs," "stay strong," "hugs," "wow," "aww," and "high five." Twitter thus asserts that the heart conveys diverse notions. Viewers are encouraged to "show how you feel without missing a beat" and understand that the heart option is intuitive and spontaneous, enables varied kinds of ideas and sentiments, and is easily integrated into online communication. The post configures what the heart means. It also indicates how the heart works as part of Twitter's interface and how members can use it. Through these frameworks, Twitter conveys popular narratives about the simplicity of buttons while distinguishing between the star and heart. Although Twitter's description of the heart button is similar to Facebook's narratives about its reactions, individuals' responses to it have been more negative.

Twitter's blogged announcement about the interface change notes that the company wanted to "make Twitter easier and more rewarding to use" and that at "times the star could be confusing, especially to newcomers."[135] Its text establishes participants as uninformed, which is the opposite of what critics of the change want. Twitter asserts that while not "everything can be your favorite" and marked by a star, the "heart, in contrast, is a universal symbol that resonates across languages, cultures, and time zones." In a similar manner, Facebook argues that the "value of reactions is that they express universal feelings that unite us all."[136] Twitter argues that the heart is more suitable for a range of emotions and forms of communication. However, before the change to the heart, the reporter Charlie Warzel described how members changed the star into "one of the most complex and cryptic forms of online communication," including such expressions as the "hate" and "flirt fav."[137] He highlights the emotive and potentially flexible experiences of the star button. Warzel asserts, "You fav because a tweet made you feel something *or* (much more interestingly) you fav because you want to make the *tweeter* feel something." He indicates that favoriting involves multiple people and systems in circuits of feeling and exchange. These reviews also suggest that favoriting can be employed, understood, and emotionally experienced in multiple ways.

Some people greeted Twitter's announcement about the heart with consternation. Such responses should have been, but were not, recognized as conveying participants' emotional sentiments. Thomas Ricker configures people's practices when reporting, "To star something is to measure its quality. To heart something is to emote it."[138] People's comprehension and valuation of these systems are supposed to be associated with "lessons learned long ago in grade school: gold stars were reserved for a 100 percent on a math test or when Mario defeated Macho Grubba, hearts were doled out like Valentines by a horny Periscope user." People's acquisition of knowledge and support of educational norms are persistent themes in literature about favoriting. Ricker uses such conventions to represent stars as appropriate, while deeming hearts as excessive, disingenuous, and related to distasteful and potentially uninvited expression. He thereby correlates buttons with gender and emotive scripts and suggests that hearts are tainting. PiotrNowinski provides a related response to Twitter's announcement by tweeting a drawing of Grumpy Cat holding an umbrella to ward off a rain of hearts.[139] PiotrNowinski suggests that the heart button is an uninvited barrage that is best kept away from bodies and identities. These participants indicate that their identity constructions and relationship with the site depend on Twitter conveying normative, and male, gender and sexual orientations.

Journalists characterize the star as more normal and reasonable than the heart through surveys and other methods. For instance, Mashable offered a poll on the "superior way to tell someone 'hey, cool tweet,'" and 82 percent indicated that the star was the best method.[140] NPR posted a survey about Twitter's change from the star to the heart, and 79 percent of respondents voted to "Bring back stars."[141] This need to numerically rank icons as methods of dismissing the heart is notable and may indicate the ways the feminine and masculine are persistently hierarchized, with the feminine deemed of a lower order and less valuable. While Twitter portrayed the heart as more nuanced and malleable, its participants dismissed its utility. For instance, jamesoreilly tweeted that the "heart is inadequate symbol. Don't want to 'heart' story about ISIS, rape, Dachau. All wrong."[142] Reporters and other participants represent the star as right and the heart as wrong and insufficient. Yet employing any star symbol as a positive method of responding to reporting on Dachau, when Nazi Germany and Nazi-occupied areas used the Star of David to mark Jews for tyrannical mistreatment and genocidal murder, is disturbing.

Reporters portray the star as more capacious and less subjective than the heart. Abhimanyu Ghoshal argues, "Unlike the heart, the star seems more neutral and open to interpretation."[143] According to Mongoosebumpkin, the "dignified, neutral gold star" is "replaced with a giggly, twee little heart."[144] Mongoosebumpkin amplifies the association of the star with neutrality by suggesting that the feminine heart is little and diminishing, and thereby threatens masculine dignity. While arguing for the interpretive flexibility of the star, such tweets refuse to interpret their own normative frameworks. Instead, they employ the kinds of emotional framework that Sarah Sobieraj and Jeffrey M. Berry outline as a method of distancing themselves from the heart and the associated references to femininity.[145] When Brownofthe-Globe tweets, "The heart connotes sentiment, the star judgment," the statement also relates the heart to unreasonable and feminine emotions.[146] Of course, such commentary demonstrates that the star and people's feelings about it are not neutral. These frameworks perpetuate gendered binary distinctions and scripts and Cartesian dualism, where society associates men with valued aspects of the mind and rationality and women with the devalued body and sentiments.

Commentaries about the Twitter heart, in a similar manner to the narratives about hearts that I discuss earlier in this chapter, reveal age biases. This occurs when CherokeeLair asserts that "hearts are cloying" and the poster is "not 13 FFS" [for fuck's sake].[147] CherokeeLair, like Mongoosebumpkin, relates the heart to youthful silliness and suggests that it demeans members. Yet CherokeeLair employs a curse that is often associated with passionate and youthful language. katecrawford supports the correlation of the heart with the cloying feminine and unpleasant residues when noting, "It's like Hello Kitty threw up in here."[148] "Not to be hyperbolic," writes Mario Aguilar, "but this 'heart' makes regular users" want to "vomit on our *keyboards*."[149] These individuals fail to analyze what the heart represents and how it works on their bodies. Instead, they reference forms of bodily disgust that are often generated when more feminine things, such as the gendered-as-female Hello Kitty and her feminine and largely women fans, are brought into correlation with purportedly advanced digital technologies.

Hearts are deemed to be risky gender scripts and to cause people, particularly men, problems by misrepresenting masculine feelings. For example, nazirology chronicles how he tweeted his "EX by clicking ♥. And she thinks" that nazirology wants "her back."[150] He concludes, "Oh wait!

#Misunderstanding #TwitterHeart." A similar position is established when ackraemer chastises Twitter because the heart is "really awkward" and makes it seem as if the individual is "sending love to random strangers."[151] Such texts convey men's unwillingness to be correlated with feelings. The associated participants worry about men's normative masculinity being damaged and in crisis because of technological and cultural changes. They represent the heart as a form of misinformation, including the ways they feel that men are improperly framed, and correlate the heart with the ways social media is employed to proliferate propaganda. Of course, Twitter has also been used as a means of undermining mainstream media reporting as "fake news" and harassing women and people of color. The hashtag sign that nazirology employs to dismiss the heart is also more generally used in hashtag hijacking and misinformation campaigns. In suggesting that the heart improperly conveys individuals' interests in romantic relationships, these people insist on defining the heart as a form of intensity that can only connote impassioned feelings. Yet people's negative responses to Twitter's shift from the star to the heart repudiate social media sentiments, even as they enact feelings in tweets and news reports.

Individuals distinguish between average Twitter participants and professional "super users," who they deem to be more important. According to emilybell, the change to the heart button "is a complete misunderstanding by @Twitter of how its super user group of sceptical journalists think of 'favourites.'"[152] And dkiesow adjusts this argument when noting that "it is not a misunderstanding but rather a repudiation of the current user base as the desired future user base."[153] Various reporters' comments suggest that journalistic practices are better than, and should be put before, other members' interests. They protest when the company shifts its gender script and associated address and tries to engage a broader array of people. This larger participant base would also presumably be the audience for journalists. Such forms of resistance to the Twitter heart, like people's refusal to consider an iPhone that accommodates fingernails, illustrate how participants patrol changes to technology scripts and protect gender conventions. Yet in critiquing and refusing the heart, journalists and other members risk making the heart into the device that they identify as endangering. I address such behaviors as a means of foregrounding and encouraging further critical interventions into the ways sites assert the identities of members and how such frameworks are taken up and refuted by individuals.

Conclusion: The Hearted Brand

Journalists understand Twitter and its features as their setting and respond to "their brand" being changed. For instance, the journalist Mathew Ingram references brand community attachments when indicating, "It's true that becoming incensed over such a trivial detail seems out of proportion in a lot of ways, but then that's exactly the kind of response that companies often see from users who are devoted to their product or service."[154] He concludes that "users care very deeply about the service and how they use it." As I have suggested in such previous texts as *Buy It Now*, brand community engagement can be profitable for companies. However, there are also risks when members' perceptions of their company do not align with management decisions. Brand community members' confrontations with products and brands include boycotts, buycotts (where individuals protest corporate practices by buying other companies' products), and emboldened critiques. In these cases, companies can experience economic and ethical challenges. The associated feelings are circulated through and by groups of people and networks when such practices are shared and reported upon. As the commentary on the Twitter heart suggests, comments contribute to the meaning of the heart and amplify the disliked connotations. Responses to the heart in other settings also employ emotive feelings as methods of rendering the site, members, and products.

Companies employ heart icons and buttons to connect individuals emotionally to their interfaces and companies. For individuals who engage in such interfaces and brands, hearts offer a ready means of expression and a sign that they are loved in response. The varied depictions of hearts that are incorporated into messages and sites configure individuals' relationships. They also allow individuals to convey their feelings for other members and experiences with products, companies, and sites. While hearts and other icons are designed to represent personalization, they are part of the generalized and duplicable aspects of sites. They are expressions that cannot be turned off or removed. Hearts are thus points of pleasure and pain that speckle every screen. They pierce people in ephemeral, unexpected, and programmed ways.

People's insistence on understanding Twitter's heart as distinct from the star button, while they are similarly employed in grade-school classes and communication, is part of the limiting gender scripts that regulate people's behaviors and visions of the self and future. Individuals who want to maintain their claims to being normative men identify the heart as endangering. Of course, the heart is an organ embedded in all individuals' bodies, as well as an emotion that is correlated with familial bonds, youthful passion, and adult relationships. When such dismissive associations are curtailed, hearts can be celebrations and exaggerations of feminine positions. They also offer women, girls, and people who identify with femininity expressive ways of seeing their interests coded into interfaces. Individuals' resistance to the heart relies on negative connotations of women, girls, and gay individuals *and* offers such cohorts opportunities for visibility and playfulness.

My considerations of hearts provide opportunities to rethink the identification of screens and buttons as flat and affectless. The online practices and critical theories that I outline in this chapter also offer ways of considering how social networks' employment of hearting and loving are supported or refused by members. Since hearts are key elements of Internet communication and demarcate the frameworks of sites and feelings, the scholarly underexamination of the functions of hearts suggests researchers' lack of interest in some of the ways gender and sexuality are structured online. There is a need for more fine-grained examinations of the varied meanings of hearts and love, which include generalized conveyances of interest, support for people and things, playful engagements, physical and emotional connections, expressions of femininity and queerness, and passionate feelings. Hearts support conceptions of authentic digital feelings and communication, as I indicate in chapter 4. As these practices demonstrate, the heart is a complex representation that underscores the entanglements of sites and feelings.

4 Screen "Tapping into your heart": Autonomous Sensory Meridian Response Videos, ASMRtists, and Tactile Addresses

Peace and Saraity ASMR describes her autonomous sensory meridian response (ASMR) YouTube video as "Tapping into your heart."[1] Since this is a tapping video in which she uses her fingers to drum on cellphone cases, Peace and Saraity ASMR suggests that her fingers, video, and devices enter into and stimulate individuals' hearts. She thus articulates a tactile version of direct address and constitutes the interests of and her relationship with viewers. I outline these haptic and emotional constructions in earlier parts of this book and later in this chapter. In this case, Peace and Saraity ASMR's narrative is infused with notions of the heart, which is associated with passionate feelings and the beating organ. The heart is also, as I suggest in chapter 3, a familiar online icon, a means of communication, and a device that is employed to stimulate emotional connections and is deemed to be too base and feminine. In a similar manner to the functions of the heart icon, Peace and Saraity ASMR touches viewers by using technologies and referencing touchscreens, and thereby she constitutes a relationship with viewers. This rapport is conveyed at the beginning of the video, where she rubs her fingers together and extends her moving fingers toward viewers in a resonant sound wave and a kind of embrace. Peace and Saraity ASMR then cups her hand to her mouth and physically and aurally extends this address by whispering, "Hello everyone."

The term "autonomous sensory meridian response" describes people's blissful reactions to auditory and visual cues, including tingling sensations. Peace and Saraity ASMR and other ASMR artists (or "ASMRtists") identify ASMR as an aspect of their practices by incorporating the acronym into their member names and indicating that they produce tingles and other forms of delight. YouTube commenters echo ASMRtists' claims and assertions of tactile artistry. For example, Be You evokes how viewers "always fall asleep to this video! It's THAT intense & amazing! The tingles" are "incredible!"[2] Be You echoes these feelings by using the heart symbol, writing, "<3 Love you girl!" Commenters describe their love for ASMRtists, ASMR videos, and sensations. Their replies are designed to heighten the work of ASMRtists. Peace and Saraity ASMR's and other ASMRtists' conveyances of tactile relationships are textually narrated, enacted through the ways they speak, emphasized by how they touch their screens (and seem to touch the associated viewers), and mirrored by viewers' responses. Emma Leigh Waldron's ASMR research indicates how "many ASMRtists strive to create perfect illusions of tactile sensation."[3] Viewers' association of aspects of ASMR videos with visceral feelings is also related to Roland Barthes's articulation of punctum. As I suggest in chapter 3, Barthes provides a theory about viewers' intense and ephemeral experiences of viewing photography that I adjust to consider other texts.[4]

In this chapter, I focus on screen-tapping videos by women ASMRtists and the ways they emotionally address viewers. Screens are a common feature of ASMR videos and are incorporated into producers' and viewers' identities and feelings. They are points of fascination and concern. For example, S. S. expresses trepidation about Peace and Saraity ASMR's screen tapping on an iPhone but "loveddd the video ♥."[5] Narratives about fragile and valued screens also appear in male-oriented technology forums, as I discuss in chapter 2. In these male-oriented sites, fingernails are often culturally devalued, identified as damaging, and mocked. While many individuals identify nails as antithetical to engagements with iPhones and other touchscreens, as I suggest in chapters 1 and 2, ASMRtists' fingernails and other feminine practices are embraced by most of their viewers.

ASMRtists, as I argue throughout this chapter, use their fingernails and other embodied parts to touch at depth and tactilely address viewers (figure 4.1). They employ their fingernails as tools to tap on objects and generate viewers' feelings. ASMRtists sometimes specify the kind of fingernails that are featured in videos and the length of these embodied features. Fingernails are also visually interesting to viewers, as I indicate in more detail later in this chapter. Viewers comment on polish applications, the length and hardness of fingernails, and the aesthetics of hand presentations. This is distinct from the cultural presumptions that I outline earlier in this book, where fingernails are presumed to constitute the abjectly feminine and to

Figure 4.1
Screenshot from Sees Nails, "ASMR Tapping On My Phone With My Long NATURAL Nails."

be associated with animalistic claws when they are long. ASMRtists' practices thus render a form of Sara Ahmed's raised arm.[6] Her feminist scholarship intervenes in the tendency of employers and patriarchal households to direct women's arms, control their bodies, and dismiss their political and aesthetic interests. Ahmed poses the willful girl and her raised arm as part of a history of feminist and labor resistance. Throughout this book, I have highlighted how certain kinds of raised fingernails are a form of feminine and feminist resistance. I continue this inquiry by demonstrating how ASMRtists reprieve women's feminine actions and celebrate the ways women employ fingernails as aesthetic objects and tools.

ASMRtists and viewers engage with each other through the actions of their hands and fingernails and the ways the associated aural, visual, and textual conventions produce individuals and relationships. They employ endearments, hearting, and what I describe as "tactile addresses" as methods of configuring and connecting with their viewers. As a means of further explaining and developing the concept of tactile address and how it is enacted through digital communication, I outline and refigure the significant literature on direct address. This literature considers how announcers and authors seem to personally engage with audience members by rendering shared settings and employing terms like "you." I also continue to develop my theoretical conceptions of digital forms of punctum, touching at depth, and to-be-touched-ness as means of considering how ASMRtists render physical addresses and intimate connections with viewers. These frameworks allow me to further consider how women ASMRtists control their position as producing subjects, which includes emotionally touching people and self-representing as to-be-touched objects.

Direct Address

Direct address, as scholars note, is often employed in television and other media forms. In my previous research, I have noted the ways direct address is used and expanded in online settings.[7] This includes the coded ability to engage viewers directly through chosen avatar names and interests. Online forms of direct address also constitute intimate bonds between people, seem to speak directly to individuals, create visual settings that continue the spaces of audience members, and parallel the positions of viewers. These online forms of address further tie individuals to settings and encourage the forms of brand love and connection that I consider in chapter 3. I consider tapping videos and ASMRtists' representations of their hands and fingernails because such formulations allow me to continue to analyze how physically touching and emotionally feeling are intermeshed. In addition, I closely read these texts as a means of thinking about the ways women's bodies are scripted and how women raise their arms to propose other forms of embodiment and experience. I find ASMRtists' tapping videos to be a useful site from which to propose a theory of tactile embodiment and address because these texts itemize, and sometimes closely read, the ways addresses function.

Direct address often represents mediated engagements as an I/you relationship and face-to-face communication. It is employed in television home improvement shows, home shopping programs, news reporting, sporting events, and talk shows. While scholarly literature usually indicates that cinema engages spectators in less direct ways than television, the feminist media studies scholar Jane Feuer describes how such filmic forms as the musical include representations of audiences and speeches from characters that acknowledge and directly address viewers.[8] Direct address encourages viewers to personally engage and feel connected to texts and products. The television scholar Michele Hilmes underscores how some forms of direct address use second-person pronouns and such catchphrases as "You deserve a break today."[9] Hilmes notes how direct addresses occur when texts, announcers, and/or characters recognize the camera and viewers. This occurs, according to John Ellis's study of media, when characters address the audience, speaking "directly 'out' of the screen as 'I' to 'you.'"[10] This acknowledgment may happen though a monologue, brief aside, laugh track, or wink that is directed at and is sometimes supposed to be produced by the audience. As my reference to the wink indicates, direct addresses are also produced by emoticon and emoji faces that "look at" and gesture to viewers. The operations of direct address may enact identity scripts and can thus perpetuate hierarchical distinctions between groups of people. Direct address, as Ellis notes, can constitute others and treat them with "patronization, hate, willful ignorance, pity, generalized concern, indifference."[11] These varied forms of direct address, as I suggest throughout this chapter, are employed and amplified online. For instance, hateful forms of direct address include Stormfront's gun-themed smilies that allow posters to virtually view and shoot at "you," as I consider in chapter 3.

Direct addresses emphasize and efface mediation. They support normative worldviews by making it appear as if viewers' physical spaces are synonymous with and operating on the same plane and temporality as represented spaces. For example, talk show couches and other representations of and references to living rooms are designed to replicate the spaces of home viewing. Matthew Lombard and Jennifer Snyder-Duch's study of advertising indicates how represented spaces also produce sensory and tactile feelings in individuals.[12] People fail to note the role of technology in rendering these settings and feelings. Direct addresses also allow media producers to render even more detailed scripts of individuals' desires, viewing behaviors, and buying habits while appearing to acknowledge personal interests. In many of these cases, direct addresses produce gender and other identity scripts about viewers, as I discuss this formation of a worldview in chapter 1 and later in this chapter.

Internet sites pick up on and sometimes restructure television narratives and use technological and programming affordances to assure individuals that companies, settings, and producers are speaking to and want to communicate directly with them. Online sites and participants appeal to, instruct, and attempt to seduce viewers. As I suggest in *The Body and the Screen*, online engagements include persistent instructional greetings to "sign in" and provide information that allows companies and sites to continue to target individuals by their names or aliases and acquire marketable information about them. There are also direct queries about people's intentions, such as "Are you sure you want to shut down your computer now?" and "Do you want to exit Netflix?" Cookies and other tracking software make it seem as if online settings have a personal connection with and knowledge of people's interests and identities. ASMRtists engage in such representations of recognition. They render direct and personal addresses by creating nicknames for viewers, promising to provide individuals with tingles, reaching out as if they are in contact, and inquiring what individuals would like to experience.

John Langer argues that viewers have an intimate relationship to television personas and devices because of the implied eye contact and other operations of direct address. He describes a "gentle yet pervasive form of paternalism where viewers, metaphorically speaking, are taken by the hand and carefully guided through the daunting, potentially chaotic complexities of life with the assurance that there will always be someone there to advise them" and "set the world right."[13] Viewers may also be taken by the eye and have their vision focused online because sites structure what viewers want to see and how they view. These acts of being taken by the hand and eye are extended and elided by computer and Internet interface handpointers, which direct individuals' attention to certain preprogrammed actions. Hand-pointers demarcate individuals' abilities to select and manipulate, but they are often coupled with set worldviews and limited options. Since they are white by default and are often associated with normative positions, as I note in the introduction, hand-pointers address white individuals and render a racial and identity script about the individuals who are expected to engage. Thus, online direct addresses, in a similar manner as television, constitute a narrow notion of users.

Direct addresses include close-up images of people squarely aligned with the frame and looking out at and seeming to mirror viewers.[14] Paul Frosh's analysis of television identifies the use of headshots, and thus faces, as methods of representing the human form.[15] The reflective screen surface also embeds individuals in televisual worlds and produces virtual versions of viewers that are supported and extended by networked computer screens. As I suggest in *The Body and the Screen* and outline in this book, mirroring representations of faces and bodies on computer and device screens turn viewers into digital images. Mirrored depictions of participants are amplified and blurred, as I describe in chapter 2, by the oily material traces of individuals imprinted on touchscreens. Even when explicit mirroring does not happen, Frosh argues that television may at any point include this "world-overlap."[16] Viewers are reminded of these unexpected moments of mirroring because people are reflected when television and other screen devices go dark. Yet networked computers and mobile devices are never fully off. Emails, text messages, and other information are persistently pushed at devices and addressed to individuals. Thus, mirroring images provide a reminder of how everyone is (or is supposed to be) always visible and trackable online.

YouTube and Direct Address

A variety of scholars suggest that people are more likely to directly address and share intimate information through the Internet than in other settings.[17] The general cultural association of online engagements with intimacy is related to ASMRtists' rendering of closeness. According to the theoretical work of Lauren Berlant and Michael Warner, intimate life is understood as a sanctuary from inequitable conditions.[18] Intimacy is a construct that normalizes how we understand bodies, relationships, and society. It relies on technologies and social structures, including gender scripts, while deploying conceptions of the natural and unmediated. Intimacy is also facilitated and mediated by conceptions of the body and skin, including the ways ASMRtists evoke touch, face viewers, and lull listeners to sleep.

YouTube renders the kind of intimate safe haven that Berlant and Warner reference. YouTube uses direct and personalized addresses and encourages the same practices from digital video producers. The site's name begins with the word "You," which suggests that the setting is for individuals and provides a form of media that is about and for viewers, who may be turned into producers. YouTube's early slogan encouraged people to "Broadcast Yourself" and to be part of the everyday structure and be visible, candid, and intimately available. YouTube's logo featured a red screen shape that framed the word "Tube," which of course is another term for "television" and references its previous production of images from a cathode-ray tube. The red, rounded rectangle is now situated before the site name and punctuated with a white arrow, which references the television remote icon and the online video option to play. This suggests that individuals should play your (and their) tube and have control over the apparatus and the associated experiences.

Menu options relate "Your videos" to the vastness of the site and company by featuring numerous thumbnails and scrollable lists of videos on the opening page and when texts are selected. Yet YouTube suggests that the interface is genuine and personally knows viewers. Indeed, the grid may have become a representation of individuals' choices since it is punctuated by options for personalization and framed by explanations of why texts are featured. For instance, YouTube's grid includes indications that blocks of text are "Recommended videos for you" and "Recommended channel for you."[19] Thus, the generic and structured aspects of the grid are remade as a singular person's array and archive that speaks directly to the individual. YouTube's and other sites' references to archives are supported by their digital organization of data, claims to comprehensiveness, references to files and folders, delivery of tagged and metadata organized content, and rendering of comprehensive versions of "you." In Kate Eichhorn's research, she correlates the archive and computer desktop. As she notes, and I elaborate upon later in this chapter, material and digital archives should also raise questions about "racialized and gendered power structures," including which individuals become visible and how they are categorized.[20]

YouTube begins to establish what it means by "you" and the terms of direct address in its "Policies" video by indicating that the site enables individuals who are geographically distant to occupy the same space.[21] The video begins with an itemization of the sorts of online content that Google, which owns and supports YouTube, delivers, and then the screen splits in half. In each frame, the viewer sees the back of an individual who is working at a computer and is a stand-in for the viewer. The people's distinct spaces are then rotated so that the individuals face each other and are situated in a shared world. Through this text, YouTube suggests that it erases computer mediation and brings addressed people into its community and a physical place.[22] In a related manner, YouTube's "About" page has claimed that it provides a "forum for people to connect."[23] Such narratives indicate that YouTube is a "participatory culture," as Jean Burgess and Joshua Green describe the platform, which is technologically driven but often identified as unmediated.[24] YouTube vloggers produce this sense of the real and of a connection through direct address when they seem to look out at, speak directly to, and touch viewers.

Direct address is a common aspect of ASMR texts, beauty tutorials, cooking demonstrations, and haul and unboxing videos where newly purchased items are displayed. Andrew Tolson's analysis of makeup video tutorials suggests that genuineness is established through "excessive direct address," "amateurishness," and the "volume and immediacy of 'conversational' responses."[25] Robert Vianello's study of live television and direct address can be updated for online sites because both forms claim that aspects of the system are live and unmediated.[26] In Emma Maguire's research on the vlogger Jenna Marbles, she asserts that people engage with YouTube because producers appear to be authentic and to share their intimate lives.[27] Vloggers' employment of stylistic conventions, which are designed to convey authenticity and naturalness, include talking about personal experiences, depicting romantic partners and friends, incorporating accidental interruptions, and vlogging from such intimate home spaces as bedrooms and bathrooms. ASMRtists employ these conventions, as well as sharing their experiences in comment sections. Viewers' expressed excitement when ASMRtists reply to commenters, as I note later in this chapter, indicate the emotional importance and hint at the economic functions of such enactments of intimacy and authenticity.

Autonomous Sensory Meridian Response

The term "autonomous sensory meridian response" is associated with a variety of YouTube videos and other texts. It describes audiovisual materials that generate pleasurable feelings or brain and spine tingles, which spread across the body. The acronym "ASMR" evokes scientific discourse but is often used in popular settings. Harry Cheadle's *VICE* article and ASMR University's online site, which seeks legitimacy for ASMR experiences through research and documentation, credit Jenn Allen with developing the term.[28] Allen developed the site asmr-research.org and uses "autonomous" to reference the "individualistic nature" of ASMR "triggers, and the capacity in many to facilitate or completely create the sensation at will."[29] She employs "meridian" as a "more polite term for 'orgasm'" and as a means of legitimizing the associated research. Yet Giulia Poerio's research on ASMR identifies SteadyHealth.com and IsItNormal.com forums, starting in 2007, as the online sites where ASMR experiences were initially considered.[30] Poerio also mentions WhisperingLife's posting of an ASMR video to YouTube in 2009.

ASMR University lists common scenarios, which are also available on YouTube, including "instructional demonstrations, methodical task completion, personal attention, focused activities, and consultations."[31] Role-play videos are common, including medical exams and salon visits. People also make videos that concentrate on ASMRtists' accents, mouth sounds, tapping, scratching, and whispering. More recently, ASMRtists have offered videos that depict the experience of being tested for and diagnosed with COVID-19. Many of these ASMR scenarios focus on touch and instances where ASMRtists' hands and bodies appear to interact with viewers. ASMRtists and their texts are thus rendered as to-be-touched-ness. Their practices are designed to activate physical and emotional responses in viewers. ASMRtists' member names, including TouchingTingles, and scenarios convey how viewers are expected to viscerally respond to touching at a distance and depth.[32] For instance, Glory ASMR describes videos where the vlogger is "touching the viewer's favce or head," even though the physical contact is acted out.[33] LJMJBI watches Chiara ASMR scratch chocolate, adopts the mirroring facets of the vlogger's direct address, and can "feel the chocolate filling the space between the finger and the nail."[34]

The writer Michael Andor Brodeur, in a similar manner to LJMJBI, underscores the intimate aspects of ASMRtists' direct addresses. He notes, "You experience an illusion of direct eye contact, a sense of proximity, a deep concern with intangible details, and a pantomime of tenderness."[35] This contact is produced when GentleWhispering strokes the camera lens as if it were your face and says, "Your skin is so soft, so silky smooth."[36] In a related manner, Jordan Pearson's article on ASMR describes when, with a "voice barely above a whisper," an ASMRtist looks into his eyes; she is "so close" that he "can hear her every tiny breath."[37] ASMRtists' indications that they know viewers and are intimately connected to them are further developed by commenters. These individuals make video requests and answer ASMRtists' queries about what kind of videos they would like to see—practices that I elaborate upon later in this chapter. ASMRtists also address viewers in such a manner that they are scripted into supporting the associated texts and the highly crafted positions.

ASMRtists' Touching and Feeling on YouTube

ASMRtists not only employ but embody terms like "touching" and "feeling." TouchingTingles's member name constitutes her identity as tactile and focused on sensations. The title of ASMR Shortbread's video "Making You Feel SO Good ♥ [ASMR] ~ hand movements, face touching" is accentuated by an image of her ecstatically tilted-back face and closed eyes.[38] ASMR Shortbread's hands are also raised and fingers poised as if in midtouch. Such finger gestures are common in ASMR images and videos, which makes the common description of these representations as "thumbnails" quite resonant. ASMRtists and viewers do not physically touch each other, but as Joceline Andersen suggests in her research, the practice of an ASMRtist "intimately connects two bodies."[39] These two bodies are expanded into a network, or what Steven Connor and Tarja Laine describe as a resonating membrane, of people who touch and are touched when individuals share and magnify their feelings in comment sections.[40]

Andersen describes ASMR as an "affective impression at a distance" and "distant intimacy" and evokes Diana Adis Tahhan's concept of touching at depth.[41] This notion of intimacy and bodies that change shape when pressed against other bodies is also related to the scholarship of Sara Ahmed and Nicolette Bragg.[42] Such connections are archived when fingerprints and other bodily traces are impressed on screens, as I note in chapter 2. The structuration of the device and the depicted individuals as to-be-touched-ness is then substantiated by these traces. In a related manner, as I suggest in more detail later in this chapter, ASMRtists promote their own version of to-be-touched-ness by referencing hands touching and rendering their physiognomy and texts as tactile surfaces. For example, Gentle Whispering ASMR evokes physical contact when she argues that viewers' sensory experiences of videos are the result of "proximity," and it is "almost uncomfortable" to be so near another individual.[43] The small size of many digital devices and screens can magnify this sense of intimate connection. Individuals have to pull images close to their faces and enlarge the representations so that they can only see fragments of depictions through their zoom finger gestures. The I/you relationship of direct address is materialized and experienced through individuals' connections with mobile devices.

Viewers are not physically near ASMRtists, but the sense of closeness is intensified by cropped depictions that represent only part of ASMRtists' bodies. This is similar to the traditional cropping and framing of women's bodies as objects and to how Laura Mulvey defines the role of women as to-be-looked-at-ness.[44] For instance, ASMRtists appear to be visually and erotically available when videos provide views that dip into their cleavage. However, individuals' desires for full views of beautiful and welcoming women are often disrupted, including when ASMRtists cannot be coaxed into including their faces in videos. Headless views risk conveying violence against women, while letting viewers heighten their relation to ASMRtists' positions. This closeness can be comforting and familiar or claustrophobic. As Gentle Whispering ASMR narrates, when viewers "feel how much" ASMRtists "care about" them "at that moment," they experience a shared temporality, the liveness of the form, and a "state of euphoria." It is at these junctures that women's position as to-be-looked-at-ness can be dissipated or combined with to-be-touched-ness.

Tasha Bjelić's study of ASMR indicates that the popularity of these videos is part of an "existing social need to be cared for, loved, and connected."[45] This care is now accessed through "digital intimacy." Andersen's ASMR research also describes a "distant intimacy that relies on the heteronormative gender roles of care."[46] ASMRtists offer videos where they appear to be tending to others, including medical role-plays and videos where they reach out as if to caress and comfort viewers, and thus script this care and availability. Bjelić's and Anderson's conceptions of digital intimacy are related to Tahhan's analysis of the ways feelings of closeness and skin-to-skin contact are experienced even when people are not physically touching.[47] Yet this should not be understood as the leveling of all difference and inequity. Touching at depth, as I suggest in the introduction and chapter 3, can also produce feelings of uncomfortableness, violation, and disgust. Bjelić highlights how people's search for online intimacy is part of "exploitative labor relationships," which include the ways ASMRtists are directed to produce texts that fulfill viewers' interests and requirements. Such requests for and enactments of care occur at a time when the state has abdicated its responsibilities to citizens and provisions for health services. Many viewers expect that women ASMRtists will care and produce videos that address viewers' desires. This includes coronavirus videos in which demonstrations of testing and assurances about contemporary experiences of distress are managed by individual women ASMRtists rather than the state.

The Feeling of Fingernails Tapping on Screens

Enthusiasts are scripted to narrate their connections with ASMRtists and pleasurable experiences with videos. This includes ASMR videos of such repetitive sounds as brushing, tapping, and whispering that conceptually bring individuals close to producers. ASMRtists structure these tapping and other sound videos to render unmediated and proximate intimacy. One of the ways ASMRtists accomplish this is by depicting their hands at the extreme "front" and lower part of the frame. ASMR Bakery takes up this position and explains that she "started tapping randomly alot, especially tapping my phone! The sound of tapping on the screen is so crisp and pleasing to the ears."[48] Viewers respond with similar accounts of their emotional interests in the sounds produced by tapping their digital devices. For instance, bee writes, "Thank you for these amazing videos," taps on her own phone in a kind of response, and indicates that ASMR Bakery's "sounds are heavenly."[49] Viewers' narratives about tapping thus connect ASMRtists and other individuals through their shared investment in sound and similar hand positions and movements. As I note in the introduction, individuals' comparable hand placements on mobile devices also render physical connections and affinities.

Yvon Bonenfant's research on sound and pleasure indicates that when individuals make sound, it creates a "resonant field of vibration that moves through matter."[50] The ways sounds shape and change flesh evoke Barthes's conception of punctum and how ephemeral viewing experiences are felt on and pierce viewers' skin. These sensations can work like chain reactions as other individuals hear, feel, and describe sounds and vibrations. In instances like Barthes's evanescent punctum, this includes the places sounds are refused or do not reach. When sounds vibrate things, according to Bonenfant, they render a form of touch. This produces a kind of touching at depth and a material intimacy because of the ways sounds echo throughout bodies and correlate participants. Bonenfant's analysis should encourage studies of the varied digital sounds that act as a form of touching, including the murmurs and pulsations of computer fans, the buzz of muted cellphones, and the vibrations of video game controllers.

Tap Asmr's member name evokes sounds and touches. She employs the phone screen to communicate with and figure viewers as part of the conversation. Her "Play games with me (nail tapping) | ASMR" video shows the

Figure 4.2
Screenshot from Tap Asmr, "Play games with me (nail tapping) | ASMR."

text messages that are directed at viewers (figure 4.2).[51] She texts, "Hello everyone" and offers a "Thank you for watching." These addresses suggest that ASMRtists and viewers are having a live, but mediated, conversation and are collaboratively engaging. Viewers see Tap Asmr produce this message as she taps the pads of her fingers and nails on the screen. They also hear the sounds generated by her interactions. Tap Asmr moves her hands to the back of the phone and connects these communications to more general ASMR tapping. She interacts with the screen and phone as a device, a processor, and a surface that delivers representations and sounds. She thus offers a compendium of the affordances of mobile phones, and presumably of how viewers engage with them. Her video and depicted phone act as a theoretical console which, according to Nanna Verhoeff, focuses people on the functions of devices and enables individuals to consider how their bodies engage with screens and other technologies.[52] Viewers are further correlated with Tap Asmr's body because the video is shot such that her hands and the phone extend away from her, and thus viewers, and into the portrayed screen space. The closely cropped images situate viewers, especially those who can match their hands to the video representations, as being conjoined with Tap Asmr and thinking through her devices and positions.

Sees Nails also addresses viewers in "ASMR Tapping On My Phone With My Long NATURAL Nails" (figure 4.1).[53] She indicates that viewers who "don't even like long nails" should "please watch" her "videos until you love long nails."[54] Sees Nails specifies that she can change the feelings of resistant viewers and identifies the persuasive and resonant qualities of her videos, which get viewers to feel the way she feels. Of course, she does not mention the amount of time or effort that such conversions will take. Instead, she gently directs viewers to adopt a more accepting position. She also correlates her texts with love and heartfelt experiences that extend beyond the general "like" structures of such digital platforms as Facebook. Ella supportively responds, "Those nails are outstanding," and notes that the sound of them makes her "ears feel nice! Thank you ♥."[55] Sees Nails replies, "Glad you liked it. Thanks for commenting." Individuals' thankfulness and gratitude are thus exchanged and amplified among participants. Such magnified emotions are embedded into the structuration of the text when Bill M writes, "That dark red polish is KILLER....♥ ♥ ♥."[56] Hearts punctuate and are employed to intensify emotional posts, as I suggest in chapter 3. By concluding posts with a "😊," Sees Nails supports presumptions about the personal aspects of face-to-face and direct engagements. In such instances, emoticons, emoji, and related icons are designed for and used by individuals to make it seem as if their bodies are placed into online settings and situated in relation to produced identities and objects.[57]

Sees Nails references connections, but in her video, her phone is turned off, which results in the reflective screen surface providing a more intense mirror image of her tapping fingernails and the room. A video fade from one fingernail tapping sequence to another multiplies the hands on the screen. At one point, Sees Nails tips the phone upward so that viewers see what looks like their own reflections on the dark screen. Such images connect the space of ASMRtists to the spaces of viewers and thereby perform the kinds of direct addresses that I outline earlier in this chapter. In the video, the ASMRtist's hands and references to touching also function as tactile addresses. Sees Nails intensifies her conflation with viewers, in a similar manner as Tap Asmr, by angling her hands into the screen space and shooting the video from above and behind her body. This results in the ASMRtist, the camera, and viewers looking at her hands from a similar position. Thus, Sees Nails delivers her identity and experiences to viewers, who see her fingernails as she sees her nails. The ASMRtist enacts her member name in a manner that incorporates the individual, who literally sees nails. Of course, viewers also look at her nails long after Sees Nails made the video because they are viewing it asynchronously.

Other ASMRtists direct viewers into the recorded setting by angling and pushing their hands into the extreme foreground of the image. For instance, thatASMRchick's hands extend beyond what the frame depicts, are elongated, and appear to reach out to viewers.[58] At one juncture, she begins to gently touch the camera, which makes it seem as if she is running her fingertips across viewers' bodies. In the last part of the video, thatASMRchick brushes the screen and waves in a manner that renders tactile connections as a form of recognition and goodbye to viewers. Such representations address viewers as the figures that ASMRtists are attending to and connecting with in physical ways. Thus, ASMRtists' representations of touch and hands are tactile addresses that act as greetings, intimacies, and articulations of bodies and identities.

ASMRtists position their hands to convey the sensations that occur when viewers are touching and touched at depth and at a distance. Thus, the hands of ASMR Shortbread and many other ASMRtists seem to hover and be midcontact in their video thumbnails.[59] The popularity of Gentle Whispering ASMR's videos, as Joshua Hudelson notes, has a "great deal to do with her hands."[60] Her "fingers arch back—rather than claw forward—as though to ensure that the contact is as light as possible." Gentle Whispering ASMR's video "•••Tapping just in Case••• ASMR Soft Spoken Gentle Tapping" starts with her touching a black surface with her illuminated hands and black nails.[61] The video contrast and different surface sheens distinguish between her matte and light hands, dark and glossy nails, and the black and slightly reflective background. Gentle Whispering ASMR's light-peach hands are visually detached from the rest of her body because the image is cropped at varied points in the video. There is also an extreme contrast between her flesh and the conjunction of nails/screen. In posing the figuration of nails/screen, I reinvoke my analysis from chapters 1 and 2 of how fingernails function as screens and forms of address.

Solfrid ASMR uses a tactile address in her many videos where she writes, "Hello my Fluffy Teddy Bears."[62] She evokes the pleasant sensations of fingers running over and grasping plush toys and connects it to the bodily feelings and textures of the audience. This greeting emphasizes the ways ASMRtists' hands and fingers work in videos and visually and emotively work on viewers. Solfrid ASMR continues to reference feelings and feeling hands when expressing her hope that individuals "like this video, and find it relaxing and tingling, and if you do, please give" her "your thumbs up and share. Thanks for watching! ˜•˜." Representations of viewers' thumbs and hands are used in this particular instance, and more generally online, as expressions of recognition and an acknowledgment that people read the text and acknowledge "you" on sites like Facebook and YouTube. Thumbs and hands express pleasure and indicate that the associated individuals like it. "Hey you" is a call that interpellates the individual under the law, as Louis Althusser suggests.[63] Thumbs-up hand icons suggest, but may not always mean, community attention, approval, and the constitution of viewers in relation to site rules. Yet thumbs also can threaten to negate some of the cohorts that YouTube and other interfaces promise to acknowledge because thumb gestures have oppressive histories. As I suggest in the introduction, opposing thumbs have been used to articulate the normatively human. The "rule of thumb" was a means of identifying acceptable tools for abusing women, and displaying thumbs in any position in the Coliseum in Rome supported the killing of gladiators.[64] Due to such associations, which still circulate in popular culture, detailed considerations of the valences and functions of thumbs-up buttons are needed.

Cultural references to fingernails and the employment of them in gender scripts, as I suggest throughout this book, also require intervention. ASMRtists offer a different framework for nails than the normative frameworks that I consider in chapters 1 and 2. They refute the association of fingernails with laziness and incapability and advertise how their fingernails function, look, and produce sounds in the titles of tapping videos. For instance, Sees Nails identifies her "Long NATURAL Nails."[65] Her use of capitalization emphasizes the materiality of her fingernails and answers viewers' questions about them. ASMRtists' titles emphasize the length of the fingernails that are producing effects and the type of nail.[66] Their identifications of fingernails as physical and aesthetic tools can intervene in people's notions of unmediated bodies and nails. ASMRtists' fingernails are described as doing and making things, and therefore being made into instruments. This happens when TheRedBaron Lives! notes that Comfy ASMR's "tapping is so good her finger nails are like a musical instrument.new subscriber earned as this was glorious!"[67] TheRedBaron Lives! thus conveys the specialized features of ASMRtists' fingernails and their creative value.

Archive of Feelings

ASMRtists' and viewers' identification of fingernails as devices, and thus as further intermeshed with screens, is influenced by the ways ASMRtists style and portray their nails. In doing this, ASMRtists incorporate their fingernails into a form of visual and aural video archive in which they itemize embodied and emotional elements. For example, Ellawyn ASMR offers "Tapping and Scratching with Long and Short Nails ASMR."[68] ASMR StitchesScritches displays a hand with long fingernails and a hand with nails trimmed to a shorter length.[69] Her fingernail actions are embedded in her member name and the practices of scritching, or making scratching motions and sounds. ASMR StitchesScritches, who is listed as StitchesScritches ASMR on Tumblr, has an imagined conversation in her FAQ in which viewers ask, "Why are your nails shorter on one hand?"[70] StitchesScritches ASMR explains that she does a "goodly bit of physical work" and needs a "hand with nails that are too short to break, can make a fist, etc." Her "right hand is not very good at doing things, so it gets to just be the pretty one." She thus displays her fingernails as a means of emphasizing the ways her hands work.

ASMR StitchesScritches echoes some people's concerns about the functionality of women's fingernails. However, she and Ellawyn ASMR complicate this by indicating that their hands have different purposes and aesthetic looks. For instance, ASMR StitchesScritches details the sounds her hands and fingernails make and emphasizes their enhanced functionality by titling a video "ASMR Sunday Shortie: lightly binaural tapping, long and short nails, fast and slow, no talking."[71] In the description, she notes that "long nails" make "tinier, higher-pitched taps" and "shorter nails" render "deeper, thuddy taps," and that each grooming strategy and body part has a purpose. Her raised and centered arms and fingernails do the sort of political work that Ahmed outlines and assert an embodiment that refutes normative cultural expectations and beliefs about symmetry. These ASMRtists thus render an archive of diverse embodied parts and sounds, including an array of fingernail lengths. This archival array is associated with the gridlike structure of YouTube, while also emphasizing mediated bodies and feelings.

ASMRtists intensify their references to the archive and archival practices in their representations of older phones. For instance, quietexperiment lists "ASMR Tapping & Scratching 6: Cell Phones of History!"[72] Her video title suggests that tapping and related sounds can provide viewers with a historical chronicle of phones. In her written description, quietexperiment directly addresses the viewer, asking, "Have you ever wanted to hear the soothing sounds of a giant gray plastic brick that also happened to make phone calls?" The individual's answer is proscribed since most viewers are looking for ASMR videos and have already read the description of this video, which says that it "contains tapping, scratching and whispering" on a "collection of cell phones ca. 1995–2012." She offers sounds and tactile explorations as a way of explaining devices. In a related manner, Cheeks ASMR provides a video in which she is "tapping/button pressing" her "old cell phones."[73] Cheeks ASMR's video causes ahmadaamer6 to itemize experiences with varied phones and to write, "The memories...."[74] Cheeks ASMR and ahmadaamer6 thereby evoke resonant experiences where pasts are produced for and shared with viewers.

ASMRtists' representations of older devices are related to the sorts of practices that the queer theorist Ann Cvetkovich considers, including her study of the emotional aspects of archiving and collecting, which are not ordinarily associated with museological and archival systems.[75] In outlining her project, Cvetkovich notes that she was interested in "how we collect feelings or store them" and in depathologizing associated forms of "loss, or mourning, as well as the impulse to hang on to things."[76] ASMRtists render a version of Cvetkovich's "archive of feelings."[77] While they do not focus on the forms of traumatic feelings that Cvetkovich emphasizes, ASMRtists queer and complicate archives through their employment of noncanonical content, use of popular platforms, emphasis on sounds and feelings as well as seeing, and evocation of nonnormative relationships and intimacies. ASMRtists also categorize their archive of feelings through varied schemas, including their identification of particular kinds of sounds, fingernails, and ASMR texts.

ASMRtists feature older mobile phones as methods of connecting people to the emotional recollections generated by older devices, the sounds that devices and ASMRtists render, and sentimental bonds and connections with vloggers. While ASMRtists' videos do not provide the mechanical information available in technology reviews and histories, they do chronicle good feelings about digital devices and the pasts that they invoke. Thus, ASMR videos about older phones directly address viewers by encouraging them to experience and virtually share specific memories about these objects.[78] They also shift the usual technological histories and frameworks of devices. Mainstream digital devices, technology archives, and gender scripts ordinarily promote the idea that designers and users of advanced technologies are male. Such structures may also emphasize the intellectual rather than corporeal features of users. Yet these technologies, in the hands of ASMRtists, are linked with the histories and archives of women's employment and interests. For example, ASMRtists include the sounds of fingernails on cellphones from varied temporal and archival moments. Thus, in Cvetkovich's terms, ASMRtists queer the "archive of feelings" by providing reconceived sound histories and new purposes for devices. This includes relating these devices to a wider range of individuals. ASMRtists also limit such reconceptualizations by associating mobile phones and other technologies with familiar memories and the related scripts.

Tactile Addresses and Responses

The archival practices of women ASMR producers bolster their relation to creativity. These women producers also emphasize their inventiveness in the many instances where they identify as ASMRtists and are recognized by commenters. Yet ASMRtists like Quiet Time ASMR retain their viewers and expand their oeuvre by indicating, "If you have any requests, then please leave them in a comment below."[79] Such notices are used by viewers as justification for any and all requests. MaricoL13 directs the ASMRtist, "Do a roleplay one where you are a teacher and you're writing on a white board or black board or even just a paper teaching something. Involve lots of paper wrinkles, pen/pencil/marker writing, tapping, stick pointing/ tracing, pages flipping, whispers ear to ear, microphone tapping with pen and such, you see the list is endless for roleplaying!"[80] While the list is a directive for the ASMRtist to move according to commands and should be considered unreasonable, MaricoL13 figures the elaborate list as helpful and instructional. In these instances, commenters' direct addresses, which are in response to ASMRtists, function as commands. These indications also support viewers' scripting of ASMRtists according to their own desires. This happens when MaricoL13 notes, "Wow a beautiful cute girl who does ASMR...girlfriend goals." MaricoL13 specifies that the ASMRtist meets the commenter's aesthetics interests and is imagined to be sexually available. These tactile addresses, which push around the hands and bodies of ASMRtists, also suggest how archives of available bodies, aural effects, and feelings are produced from ASMRtists' labor.

Some commenters' lists of interests constitute the bodies of ASMRtists and the more general bodies of women. Commenters' "requests" thereby function as gender scripts. This occurs when Alladin Zilva provides a detailed script of desired scenarios that features the ASMRtist's nails but frames it as a request: "pls.do videos like flipping pages, showing things in handbag, writing, putting on make up etc.. with bright colors of ur nails. Love to see ur sexy talons in action."[81] In a related manner, Stephen Arcella conveys expectations that the ASMRtist will be available to serve his interests. He writes, "Please do another video showing your gorgeous nails with clear polish on them. Would you show more of the undersides of your nails, especially your thumb nails?" He would also like "a measuring video" because he is "very curious to know exactly how long" her "beautiful nails are."[82] Arcella provides a detailed list of how the ASMRtist should chronicle and measure her body for viewers. While prefaced with a version of "please," there is a presumption that viewers' positions as directors and their interests are central to ASMR practices and the associated videos. These comments are related to individuals' sexualization of the bodies of ASMRtists, which I consider in more detail next. The requests are often accompanied by ASMRtists' indications that they will try to provide such texts, or that they have already produced a similar video, and thus that the viewers' interests are part of their YouTube archives. For instance, alsrg asks for a video, and Chiara ASMR responds that she has "done one :)" called "one hour of tapping!!"[83] While ASMRtists directly address viewers, these commenters indicate the ways viewers direct women vloggers and figure them as objects of contemplation.

Viewers' requests allow ASMRtists to render their videos as ongoing conversations. Thus, mediated addresses seem to be, and for some viewers are, dialogues between participants. These arrangements are structured by the YouTube format, with its options for producer and viewer commentaries, and by the ways ASMRtists arrange request videos. ASMRtists employ the YouTube interface to extend the implied interconnections and directives of request videos. For instance, ibokki ASMR creates a video titled "LEAVE

YOUR ASMR REQUESTS," which functions as ASMR and is designed to generate more ideas, engagements, and viewers.[84] The comments with the "highest number of likes" or that she thinks "are just awesome" will be made into ASMR videos. ibokki ASMR figures the community's interests while suggesting that she can recognize any individual for a remarkable request and transform it into content.

ASMR videos function as mixtapes of members' interests and represent viewers as authors. This happens when Clareee ASMR provides "ASMR Doing YOUR Requests!! –Assortment" that she garnered from a YouTube poll.[85] Gibi ASMR includes a related compendium of "[ASMR] Top Requested Triggers ~."[86] These ASMRtists structure feelings through consensus and numerous requests. ASMRtists' response videos are a compendium of viewers' interests and a representation of the conversations and interests that they share. Cvetkovich's identification of mixtapes as an archive of feelings applies to these practices. They are "evidence of how friendship and the connections we make" matter.[87] The corpus of ASMR videos also functions as an enormous mixtape and an archive that matters, especially when such options as autoplay are enabled and the next video automatically loads and then screens.

In the case of ASMR videos, commenters may also try to produce matter in the form of gender scripts about how women are expected to render their bodies. ASMR response videos are tangible records of such engagements. These activities may be amplified, as I suggest in the introduction and chapter 2, by the ways people's fingerprints and other movements appear on touchscreens and their hands rest in similar positions. ASMR requests generate a visceral audience and set of interests for queries. They result in a higher likelihood that ASMRtists will attract viewers. They revision ASMRtists' claim to be artists, who are ordinarily expected to work individually and to have control over their artistic output into community and collaborative production (and possible oppression). Since ASMR request forms generate reflexive lists of ASMR practices and ideas, they offer the opportunity to contemplate the functions of ASMR and the related request practices that occur in different online production communities. Requests continue to be an aspect of fan vidding where clips from favorite shows are edited into new configurations and storylines, avatar production, and other kinds of online production where these practices place participants in conversation and render mixtape compendiums of authors and texts.

Intimate Responses

ASMRtists' video descriptions and queries about requests are formulated to generate impassioned responses. YouTube's design, including its "Comments" section, is intended to produce such engagements. ASMRtists follow and encourage the interface's scripts about genuineness and intimacy when they punctuate comment sections with hearts and other emotional responses. For instance, Peace and Saraity ASMR enacts versions of the emotive connections when describing her video as "Tapping into your heart."[88] In reaction to Peace and Saraity ASMR, Jordan Maxwell uses capitalization, which is understood to represent intense feelings and shouting, writing, "OMG SHE LIKED MY COMMENT."[89] The commenter feels "SPECIAL" and asks for her to "REPLY PLZZ." When ASMR Bakery responds to a post, Emma E emphasizes her excitement when writing, "Holy crap she replied."[90] Maxwell and Emma E indicate that the feelings generated by ASMR videos are connected to and intensified by ASMRtists' personal attention.

Maxwell's and Emma E's responses may be particularly apt since ASMR videos are devised to convey personal contact and care. ASMR replies produce an expansive and nuanced representation of intensity and love through the employment of repetitive and multiple kinds of hearts. For example, Kayla Livingston's indication that she loves Peace and Saraity ASMR's makeup is amplified by a series of smiling faces with heart-eyes.[91] Peace and Saraity ASMR hearts the post, and Livingston responds, "omg you replied" and includes a red heart. These practices, which fill some comments sections with hearts and related emotional icons and texts, convey participants expectations about and the recurring rituals of responding in a setting that emphasizes uniqueness and authenticity. Since heart emoticons and related expressions become an archive and language of feeling, which is persistently drawn upon, these conveyances and punctum may also flatten the associated responses and the influence of these frameworks.

Commenters' descriptions of their feelings about videos are also often comments about ASMRtists' embodied features. Such commenters presume an intimacy that enables them to detail other people's bodies. For instance, Mike Sanchez comments, "Love the pink nail polish and shiny top coat! Love the length and shape of your gorgeous nails!! And the close up of them, is breathtaking!!"[92] He goes on to expound on how the surface, contour, and video rendering of the ASMRtist's fingernails makes him feel. In such instances, commenters support notions that the hands and fingernails of ASMRtists, which are focused on producing and representing sensations, have emotionally influenced them. ASMRtists' representations of their bodies and commenters' narratives attempt to change produced texts into unmediated personal intimacies.

Commenters adapt ASMRtists' intimate addresses as methods of conceptualizing further contact with the ASMRtists. For instance, Daryk fancies being in physical contact and asks other readers, "Imagine getting your back scratched with dem nails."[93] Aero renders a more misogynistic framework when identifying a desire for the ASMRtist to "scratch my balls."[94] Rather than solely figuring ASMRtists as available to touch, and thus constituted as to-be-touched-ness, these individuals direct ASMRtists to touch them. This is not inherently an empowered position for ASMRtists since commenters adapt ASMRtists' construction of tactile worlds and take control of the associated narratives and bodies. Such erotic narratives may unfortunately seem warranted because of the ways ASMRtists address and make themselves available to viewers. This is compounded by many ASMRtists incorporation of requests into their practices.

ASMRtists tend to portray videos and the associated structures as intimate conversations and requests. More than expressing the ways they feel about the videos and the associated ASMRtists' bodies, this encourages some viewers to try and correct ASMRtists' self-presentations. For instance, a poster uses the member name and phrase "careful aesthetic" to call for normative self-presentation, offers a prescriptive lesson when writing, "See girls.. (and guys) These are good nails. Don't get fake acrylic ones that you can't type, text, or even touch correctly with! – Be Natural."[95] Such comments render gender scripts about women (and less frequently men), fingernails, and device use. The associated commenters mandate women to present a natural physiognomy and at the same time contrarily inform them that they need to correct their embodiment.

Charlotte delivers and simultaneously revokes a compliment when noting, "Love your vids. Your makeup is on point always but gurrrl those nails are trash."[96] Her use of the term "trash," in a similar manner to concerns about the relationship between hands and animal parts, removes the associated ASMRtist from the classed and human. Charlotte continues to articulate visual displeasure and the abject and inhuman body when she claims that she "couldn't watch the videoooo." When Jairo L. is chastised for directing the ASMRtist to "cut that nasty nails (no offense)," the commenter claims to be "doing her a favor, if no one says that to her she will never realize that her nails looks disgusting."[97] Since ASMR is focused on pleasurable and persistent looking and listening, these warnings about unwatchable bodies work to remove the vlogger from the subject area and the category of the human. They render ASMRtists as feminine figures that are in need of regulation and advice about how to be desirable and intimate objects. These commenters provide unasked-for critiques about ASMRtists' bodies—a practice that is familiar to women—and argue that deprecation is a form of care. In making these arguments that love and intimacy are conveyed through negative commentary, these posters risk justifying verbal and physical violence against women. Yet ASMRtists also push back against such objectification by consistently foregrounding their active hands.

Conclusion: Making ASMRtists and Their Hands into Subjects

ASMRtists' and other producers' direct addresses, smiley speckled responses, and video close-ups are designed to connect the faces of speakers to the faces of listeners. ASMRtists' screen-tapping videos also link the fingers of the tapping ASMRtist to the digits of viewers. ASMRtists' representations of mediated fingers, which are sometimes scripted as touching "through" the screen, emphasize the ways digital technologies convey embodied digits and feelings. As I have suggested earlier in this book and in other texts, this enmeshes finger digits with digital technologies as a means of extending and articulating the body in online settings. Ayla ASMR underscores this through the title of her video, which is "ASMR ≈ ♥ Delicate Tapping ♥ iPad and iPhone || AylaASMR."[98] She connects gentle finger taps and the sounds produced by touching digital devices to heart icons and heartfelt feelings. The hearts act as a sign of her taps and perform as a version of Barthes's punctum, or as tender points.[99] In a related manner, the Unicode addition of a "Heart Hands" emoji in 2021, which depicts two hands touching in such a manner that they form a heart, suggests that haptic contact produces hearts, love, and connections between people.[100] Hearts are also punctuation, including exclamation marks, producing and underscoring pulsing and affective resonance. They connect people's resonating heart membranes to the feelings that are mapped onto other embodied and technological parts.

Ayla ASMR asks viewers to "OPEN ME." Her phrase refers to the You-Tube option to "SHOW MORE" of vloggers' narratives. Her text suggests that she and participants should open themselves up to and be available for exchanges of feelings. She guides them to respond with their own hands and offer a "Thumbs up if you also love tapping and catch yourself tapping on your phone!" Ayla ASMR's and Solfrid ASMR's indications that viewers should make their own finger gestures through YouTube's "like" option are related to, but not the same as, the fingers that are engaged in delicate tapping.[101] Ayla ASMR's viewers take her up on this indication about loving to tap. ASMRtists' and viewers' mirrored narratives about tapping allow them to render shared interests and feelings. This conception of sharing is fostered by the ways direct addresses connect the spaces and values of participants. For instance, goodpigASMR, who has a tapping video, replies that she taps her "phone all the time!!"[102] Ayla ASMR replies to such chronicles of viewer tapping and notes that she is "not alone!!! #tappersunite."[103] These individuals constitute shared times and suggest that they were and are tapping in unison.

Many ASMRtists' tapping videos, as I demonstrate throughout this chapter, provide positive demonstrations of women's working hands and fingernails. Their commenters often support these positions. For instance, Eden Cameron indicates that an ASMRtist has the "best nails for asmr."[104] The visual features of ASMR and women's bodies are further emphasized when skinny legend writes, "your nails are visually pleasing,"[105] This lure of visual pleasure, which can fix the viewer in front of the screen at the same time as it constructs ASMRtists, is also conveyed when BTS Trash rhetorically wonders why the commenter is "watching so much of this? Your nails are so pretty and the sound is so soothing."[106] ASMRtists' fingernails thus function as visual devices as well as tools. These represented body parts tactilely address viewers. I employ the literature on direct address as a method of developing ways of considering how such acknowledgments function in ASMR videos and other online sites. Digital forms of direct address are conveyed through texts and interfaces that claim to recognize individuals.

As I argue in this chapter, ASMRtists' tactile addresses and related practices render positive representations of women's hands and the utility of their fingernails. This includes correlating women's fingernails with touchscreens. This is distinct from dismissals of women's interrogations of iPhones and the ways these devices do not work with fingernails. The participants in technology forums also tend to denigrate women with fingernails. Nails are connected to and embedded in the human hand but are perceived by some as changing individuals' fine-tuned instruments into clumsy and brutal animal claws. Such associations are more likely to target women and people of color and thus have misogynistic and racist connotations. People's concerns about dirt and contamination, as suggested by my considerations in chapter 2 and the afterword, also tend to be directed at hands, especially the hands of already dismissed subjects, since they are used to eat and to touch other people.

Parents and individuals performing elder care frequently focus on the cleanliness of their charges' hands and fingernails. Parents often ask, "show me your hands" and "show me your nails" when looking for concealed dirt and other transgressions. "Show me your nails" is also an explicit or implicit request directed at nail bloggers and nail artists because of their skillful designs. Given the use of the phrase by parents as a patrolling gesture and as a means of playing with and distinguishing different parts of the body (hands, nose, toes...), the use of "show me your hands" by police is an extension and negotiation of familial and state control. This address is a method through which individuals are interpellated by the law, as theorized by Louis Althusser, and made into errant, constrained, and endangered subjects.[107] Yet fingernails are also part of ASMRtists' creative practices and a site where they highlight their tapping tools and visual features, which I consider in more detail in the afterword. In these cases, the "hey you" of subject formation and direct addresses is designed to produce connections between ASMRtists and viewers and heartfelt sentiments, which may still constrain aspects of how women can engage.

Individuals employ tactile direct addresses to amplify people's feelings. Tactile addresses shape the ways people engage and experience a variety of online practices, including ASMR and ecommerce. I thus believe that studies of tactile direct addresses can advance the research on affect and touching. This includes understanding how direct address functions, screens are rendered, and viewers are constituted, recognized, and made to feel. Therefore, scholars of digital media and popular culture should closely study how individuals are addressed with feeling. This involves how individuals engage with and characterize others, including the ways hate is amplified online through the ability to communicate with like-minded people, render other groups in straw form, spread misinformation, and convey and experience magnified feelings. Screen cultures may appear to have displaced or downplayed tactility. However, physically touching and emotionally feeling are ever referenced and key factors in the ways we view and are understood. Whether it is representations of buttons or scratches on a screen, tactility is meshed with and sometimes disrupts digital visual representations.

Afterword   Being "less touchy-feely" During the Pandemic: Socially Distancing and Emotionally Feeling

Squeaker Tweeker employs Urban Dictionary to define "Social distancing anxiety."[1] The poster chronicles the "palpable anxiety" that occurs when considering "being separated" from "fellow humans, even at six feet." According to Squeaker Tweeker, social distancing anxiety includes "Fearing" that the "State will lock" the poster up alone for a long period of time. These comprehensible sentiments are amplified by Squeaker Tweeker's member name, which references someone who grapples with addiction, fear, and distrust. Of course, domestic lockdowns have become a part of many individuals' lives because of governmental attempts to manage the coronavirus pandemic. As a means of conveying these kinds of experiences, Squeaker Tweeker includes a Graphics Interchange Format (GIF) file that illustrates the described confinement. It depicts a single skeleton standing behind a window and waving desolately. The skeleton's loss of flesh and the surrounding empty windows emphasize embodied dissipation. The definition and GIF suggest that individuals and bodies are formed through contact and connection with other people so that the isolated individual is socially and corporeally amorphous. Through such formulations, Squeaker Tweeker highlights how socially distancing is connected to emotionally feeling.

Medical professionals and journalists have noted how the coronavirus pandemic has heightened people's anxiety and depression.[2] This literature has more generally underlined an array of feelings. While Squeaker Tweeker values contact, the psychologist John M. Grohol's article "Coronavirus Anxiety: Social Distancing Helps Stop the Spread" suggests that people will feel less apprehensive when maintaining social distancing.[3] Physical distancing, according to the Centers for Disease Control and Prevention (CDC), is a synonym for social distancing and a method of "keeping a safe space between yourself and other people who are not from your household."[4] It is thus also a schema where individuals articulate relationality and risk. Social distancing is conceptualized as a kind of body associated with corporeality and defined as embodied measurements when the CDC identifies it as being "6 feet (about 2 arms' length)" apart. Thus, the CDC and others continue to associate being apart with bodies, including feet and arms, that they want to avoid contact with and remain distant from. Of course, Squeaker Tweeker and others suggest that safe spaces are not identical with spatial distance or domesticity. The safety measures and rituals that have been developed to address the ways the pandemic threatens people's lives, long-term health, and livelihood have also altered (at least temporarily) some people's understandings of embodiment, touching, relationships, and home.

The reporter Ali Pattillo provides a precoronavirus history for social distancing. Before 2020, writes Pattillo, "social distancing was just another phrase for avoiding your ex or an overzealous coworker. But this year, as the novel coronavirus spread, social distancing became an essential part of our lexicon and one of the best strategies for staying healthy."[5] She relates the past of social distancing to relationship avoidance. This perceived shift is also from social to embodied avoidance. Social distancing is conceptually rethought, according to Pattillo, so that it is no longer, or not solely, a social tactic; it is now a strategy for staying alive. Of course, narratives about social distancing and physical contact become enmeshed with conceptions of estrangement, emotionlessness, and comfort. Social distancing also evokes ideas about closeness and affection that sometimes are the outcome of caring to keep people safe and at other times are what the virus and social pressures have revoked. People experience social distancing as deeply emotional when fighting over expectations that individuals should follow CDC and other guidelines and asserting that such parameters are unnecessary.[6] Physical representations of social distancing have been correlated with emotions. As Vittoria Traverso reports, the use of chalked circles in parks to designate individual and group boundaries have been identified as a safe and comforting shape.[7] Such markers also convey a cultural schema that keeps people from individually having to assert their own expectations about distance.

People's Urban Dictionary definitions of "social distancing" encapsulate this oscillation between intimate contact and personal protection. According to Gartholomew83, social distancing is what "you claim to do when someone is into you, but you want to avoid them and not be rude."[8] Kaydog1 elaborates upon a more personal refusal when associating it with what Kay-dog1's "wife practises in bed." This poster chronicles being *"up for sex"* but *"she got as far away as possible, called it social distancing,* thereby combining gender stereotypes about women's purported indifference to sex with pandemic health practices."[9] This allows Kaydog1 to play with but also undermine the reasons that women seek sexual agency and the conditions that encourage people to consider the health of individuals and communities. NoThankYou4PiggyFlu's member name is also about refusal; it alludes to avoiding contagion because of the swine flu. This person defines social distancing as a "technique used to prevent the spreading of a pandemic disease by physically distancing yourself" and as "code for private, solitary activities of *Any* kind. (i.e. masturbation)."[10] Here too, social distancing becomes a way to frame sex. Social distancing is further intermeshed with cultural notions of failed relationality when NoThankYou4PiggyFlu provides the example of an individual who is *"going to go do some heavy Social Distancing!"* A friend is imagined replying, *"You need a girlfriend, man."*[11] Women are thus figured as necessary to men's normative identity and embodiment and as ruptures to this system. Individuals on Urban Dictionary outline the ways social distancing acts as a framework for moderating contemporary relationality *and* how it causes uneasy feelings.

Chris Stokel-Walker reports on how "personal contact will change post-Covid-19."[12] The journalist argues that people will "be less touchy-feely and far more wary, but the transition will feel strange." Such formulations ordinarily indicate that cultural changes will be accepted, or even embraced. However, he uses "but" as a qualifier to note that the change is going to be emotionally and physically difficult. Stokel-Walker employs multiple emotional and tactile terms to narrate how people are emotionally managing fluctuations in connecting. People are expected to "be less touchy-feely" and to "feel strange" about touching. Thus, feeling queasy is forecast to prevent people from feeling and to make them experience emotional distress. These frameworks contrarily dismiss and escalate the links between people's cultural functioning and feelings. Such chronicles about emotions, as well as indications that there will be significant changes in people's sentiments and tactile engagements, are integral to pandemic journalism and what some people identify as "post-pandemic" reporting. Journalists' and other people's accounts thus underscore the ongoing intermeshing of physically touching and emotionally feeling. They also indicate current concerns about people's experiences and that the dissipation of touching, including reporting on these issues, is filled with feelings.

The queer theorist Eve Kosofsky Sedgwick, as I suggest earlier in this book, emphasizes the ways individuals and cultural structures correlate physically touching and emotionally feeling. She highlights the connection between sensations and the trivialization of the term "touchy-feely." In a slightly less dismissive register, as Sedgwick notes, references to touchy-feely also imply that even talking about "affect virtually amounts to cutaneous contact."[13] Stokel-Walker's narrative about the term "touchy-feely" suggests that people were wary about such tactile and emotional bonds even before these sensations were recontextualized by the risks associated with the coronavirus pandemic. People's terminologically highlighted wariness about excessive and hyphenated touching and feeling may dismissively associate such practices with the feminine and queer. Yet popular literature, including Stokel-Walker's article, and psychological and developmental research underscore that individuals, including children and the elderly, suffer emotional, developmental, and health consequences when they do not experience physical contact.[14] My study of journalists' coronavirus reporting also notes that women are frequently the producers of such accounts and the news media often associates women with lifestyle reporting.

These reports and some of the other practices that I outline in this book, including the employment of heart icons and production of autonomous sensory meridian response (ASMR) videos, are associated with women and femininity. They are designed to replicate and expand people's interests in experiencing pleasant feelings. Such practices work through and are undermined by the screen-based aspects of these engagements. Correlating the screen with the body helps to resolve such quandaries and to maintain people's investment in materiality and authenticity. Throughout the previous chapters, I have thus considered the ways the body is made into and understood as a screen, is supposed to be in physical contact with other bodies through screens, and is engaged in touching and feeling. Bodies are rendered as screens when they convey feelings through facial expressions, hide devices, and are connected to and act as frames for screen-based technologies. Screens are also revisioned as bodies when they are depicted as living and dying, wrapped in skins, and identified as intimates. For instance, ASMR producers tactilely address viewers by touching screens and cameras. Even as the pandemic persists, ASMRtists have focused on physical contact

and attended to viewers' feelings. Yet ASMRtists' and other people's representations of embodied screens and practices of touching and sharing
screen surfaces have been met with increasing consternation because of the
coronavirus pandemic.

I analyze some of the ways touching and feeling continue to be intermeshed, even as people worry over the viral implications of contact. I focus
on ASMRtists' and journalists' articulations of touch. Their accounts of
physically touching and emotionally feeling, especially when correlated
with a pandemic, foreground the mixed emotions that are associated with
gendered, raced, and sexualized subjects. ASMRtists' and journalists' narratives also allow me to consider the ways the cultural association of physically touching and emotionally feeling persists and is hinged to socially
distancing and emotionally feeling (and being identified as unfeeling).
This occurs along with worries about infection and dismissal of such concerns. For instance, Eric T. Lehman's study of touch and music during the
pandemic considers how Neil Diamond revised the song lyrics for "Sweet
Caroline" as a means of referencing handwashing and contact avoidance.[15]
Lehman argues that Diamond's and other individuals' denouncements of
touch are cause for concern because touch is essential to people communicating and sharing feelings. My proposals for touchscreen theories offer
methods of addressing the cultural and personal implications of such structures. I also provide ways of thinking about how people enact and write
about physically touching and emotionally feeling while these experiences
are longed for and challenged.

Touch, as my previous chapters of this book indicate, may be identified
as a key aspect of human connection and communication, but it is also
associated with an array of binaries, including being pleasant and unpleasant, clean and dirty, and welcome and unwelcome. It is also linked to rela-
tionalities and feelings. The more negative understandings of touch have
been present over long periods of time and, as Lehman notes, there is a
history of touch being "conflated with disease and contagion."[16] In Martin S. Pernick's consideration of contagion, he outlines how the term is
"derived from the same Latin root as *contiguous*, meaning 'touching.' Thus,
in ancient and medieval medicine, a contagious disease meant one that
spread from person to person by touch."[17] In recent writings, the correlation of contagion and touch has persisted even as the research and reporting about the coronavirus and associated COVID-19 illnesses indicate that

spread tends to occur through expelled droplets and smaller, aerosolized particles that include viral materials. Contact with the other continues to be demarcated in relation to material and coherent bodies rather than the more amorphous and open ways bodily matter influences and drifts through settings. This breath and aerosolization can be critically considered through Diana Adis Tahhan's theory of touching at depth, rather than the more delimited notion of touching as condensed in the hand and a small number of other body parts.[18]

I argue that popular writing about the coronavirus pandemic has emphasized the relation between physically touching and emotionally feeling *and* socially distancing and emotionally feeling and not feeling. These evocations of not feeling include concerns that people are no longer emotionally engaging with other people, indications that digital tools have replaced material connections, worries about diminishing interpersonal touch because of social distancing, and dismissals of women's agency and feelings about unwelcome touching. Reporters and related individuals tend to correlate the concept of not feeling with feeling, all while invoking gender norms. They chronicle and record emotions when considering social distancing because of the coronavirus pandemic and the history of digital and tactile connections. There has also been a cultural tendency to reassert norms in people's keen (but sadly premature) celebration of the end of the pandemic in parts of North America and Europe during summer 2021 (and more recently). This includes journalists and the public's uncritical adoption of narratives about the good feelings, and touch, that are associated with a return to "normal." Jubilant reporting on the return to "normal" has failed to address systemic disenfranchisement and critiques of norms.[19] It has instead emphasized people's increasing socialization and bodily contact. This journalism has imagined a contemporary moment and future in which people can more directly address each other because they are not screened by masks, online communications, and digital devices.

ASMRtists' Feelings

ASMRtists' screen-tapping videos, as I suggest in chapter 4, take a different approach by foregrounding and eliding digital mediation. They portray active hands and fingers as a means of tactilely addressing viewers. ASMRtists intend for these representations to facilitate pleasant, or "tingling," sensations. Their coronavirus videos, which I outline later in this afterword, continue to emphasize producers' tactile care work and address viewers. ASMRtists inform viewers that their videos provide ways of grappling with stress and concerns about touching that are associated with the pandemic. For instance, Meditative Lullaby ASMR offers a "soft-spoken ASMR-style meditation with the aim of easing your anxiety about coronavirus."[20] She "will also be simulating face touching, creating hand sounds, and plenty of visual hand movements in an attempt to provide you with a deep sense of relaxation." Meditative Lullaby ASMR references intimacy and a kind of gentle aural touching through her reference to "soft" speaking, which is designed to make people feel better. She also foregrounds the simulated aspects of embodied contact, which are correlated with viewers' sensations. In a related manner, news coverage about the virus interconnects physically touching and emotionally feeling. Journalists, as I suggest in more detail later in this chapter, highlight and mourn the contemporary displacement of embodied contact for health reasons, including tactile connections with digital devices. Yet ASMRtists continue to evoke physical connections between bodies, cameras, and screens because of their interests in emotionally experiencing textured and sounding surfaces. They also articulate their control and virtually touch the coronavirus when performing as laboratory test takers, diagnosticians, and other healthcare workers. They provide video sequences where they seem to eat the virus.

Near the beginning of "ASMR Calming You Down (corona virus edition)," Ozley ASMR extends her hands toward the camera, screen, and individual.[21] Her movements are designed to connect with individuals by alternately touching her hair and face, caressing the fur-covered microphone, and reaching toward the foreground. She therefore appears to be closer to viewers. Madison Phoenix ASMR's video for "Coronavirus Anxiety Relief" also features her reaching out to viewers as a means of offering a face massage.[22] Her face, which is often the focus of the video, remains more clearly portrayed. This results in her hands blurring and seeming to become soft as they move closer to the camera and viewers' screens. The video's visual, tactile, and auditory elements, including her whispering of "so close, so close," support her rendering of intimate contact. Madison Phoenix ASMR's use of the term "close" sustains her production of an intimate space that is shared by the ASMRtist and viewers. The term "close" enacts, and may magnify because of the evocation of spatial nearness, a similar series of physical and emotional intimacies. Her articulation of being close thus intermeshes the experiences of physically touching and emotionally feeling. This is designed to produce a form of Tahhan's touching at depth, where emotional experiences of connection occur at a distance but feel intimate. It also produces a tactile address, which I detail in chapter 4 and later in this afterword, and addresses viewers by seeming to enwrap them in sensations.

Madison Phoenix ASMR enacts a series of typical ASMR practices when she features the microphone, touches and taps it throughout the video, and portrays the ways she produces sounds. In Gibi ASMR's video description, she expressively notes, "Ahhh, camera lens tapping!!!!!"[23] It is one of her "favorite triggers in videos that is usually unintentional/meant to be 'touching you' during an exam, makeup, etc." As she suggests, when ASMRtists reach out and tap the camera, microphone, and screen, they appear to tap and touch viewers and render personal engagements. Gibi ASMR foregrounds related forms of emotional and tactile direct addresses. This includes portraying textured connections, sharing expressive sentiments, seeming to tap and breathe near viewers, and using such terms as "you" to appear to be personally communicating with individuals.

ASMR viewers ordinarily engage with breath and touch as delightful sounds and intimacies, but these experiences have more recently been rendered as endangering because of the coronavirus. In Bunny Marthy's ASMR video, she picks up on the experiences of pandemic mask-wearing by breathing heavily behind a pollution mask, which is designed to protect individuals from toxic particles, and recording the auditory results.[24] Marthy renders her body as feminine too much and too close by doing this. Gibi ASMR apologizes for being "in your personal space," which is another form of shared and direct address, and then asks if it is okay to "touch your face." While this notion of intimacy is welcomed by many viewers, Perry The platypus writes, "Don't touch" because "it's corona time."[25] Such comments point to a quandary for ASMRtists because their videos are ordinarily about touching. Nevertheless, a group of ASMRtists continue to engage in touching at depth where the sensations of contact do not necessitate physical proximity.

ASMR videos encourage individuals to experience their bodies as connected to and made over into technologies, in addition to being linked to ASMRtists. For instance, Halunke Nr. Eins identifies the "intense feeling of being a camera" because of Gibi ASMR's video.[26] Seafoam Kitten's ASMR conveys a similar personal and technological metamorphosis in "ASMR - YOU ARE MY CAMERA ~ Personal Attention and Affection! Lens Touching, Button Tapping ~."[27] These ASMR videos render, at least for some, a cyborgian meshing with ASMRtists and devices and reconfigured ways of understanding the self. Gibi ASMR articulates how she meshes with viewers and devices when describing her fingernails as "tapping tools" that touch and make sounds on surfaces and individuals.[28] The journalists Geoffrey A. Fowler and Heather Kelly engage in the similar reshaping of objects and contexts when characterizing general screen engagements as a "survival tool."[29]

ASMR viewers relate becoming and existing as cameras back to being bodies by chronicling their experiences getting touched and changed at a distance and depth by ASMRtists. For instance, speed wagon imagines a visit to the doctor after being in the position of the tapped and scratched camera. speed wagon describes "intense scratches" because the ASMRtist "attacked" the viewer's face with "her tapping tools."[30] Louise Anne A. identifies a point in the video where she is "stabbing your eyes and face with her tapping tools."[31] Such comments underscore the visceral experiences of ASMR videos, the ways viewers pick up and support the language and conceptions of ASMRtists, and how such corporeal experiences are explored and embraced. Thus, such texts offer ways of not only intensifying bodily experiences, but articulating and developing the body's association with technologies and mediation.

Researchers could further consider how ASMRtists' and viewers' imaginative accounts of being media technologies are enacted in other YouTube practices and online forums. ASMRtists' technologization of the body as camera and screen are intended to provide some relief from concerns about the kinds of embodied precarity that have been intensified by the coronavirus pandemic. ASMRtists also render more challenging bodily encounters that viewers interpret as physical damage to their bodies. These shifts in context are underscored when Spitfire11511 asks who is "watching this in 2020" and directs Gibi ASMR to "get ur hands away."[32] Aldo gamas just writes "Corona," presuming that no other comment is necessary for articulating the seemingly embodied and hazardous relationship between ASMRtists and viewers.[33] These viewers' comments underscore the visceral intimacy and sense of closeness rendered by ASMR video producers.

ASMRtists' rendering of calming and pleasant sensations has continued and been changed by viewers' indications that ASMR is linked to the coronavirus. Stanley Mikko Santiago asks, "is asmr infected by corona virus?"[34] This suggests how ASMRtists' focus on contemporary feelings and conditions is identified as potentially contaminating. Sc O proposes related conjunctions of appeal and disgust when noting, "Your mic is a fluffy corona virus."[35] ASMR ordinarily enlivens the surface of videos and associated technologies, but in this case, it also animates the invisible but persistently illustrated shape of the virus. ASMRtists resist the pandemic by engaging in such practices as eating and thereby abolishing the virus. This is not surprising since ASMR is often constituted as curative. For instance, Tim lovely identifies whispering as "so damn incredibly relaxing that it could become a cure for that virus."[36] Gerard Zandvliet foregrounds his attachment when writing, "Your videos will be a tremendous help" to "stay positive and not feel lonely."[37] The associated producers and videos render a sense of closeness and intimacy that is not reliant on physical contact. ASMRtists and viewers thus constitute a kind of connective tissue, including waves of sound and hearts that punctuate descriptions and bodies.

Viewers use visual devices to convey how their pandemic experiences are improved by ASMRtists. For instance, they employ heart icons to recognize how ASMRtists provide comfort. ASMRtists reply with similar forms of hearts and thereby work to circulate the kinds of sentiments and hearting that I consider in chapters 3 and 4. ASMR SWEETIE repeats and intensifies these ideas of connection, which take on a kind of virality through their multiplication, by writing, "you're not alone! We're all in this together and we have to find comfort in that, or at least try."[38] She concludes by expressing her "love" and punctuating and amplifying her message with a series of hearts. HeatheredEffect ASMR similarly assures viewers, "You are not alone if you feel sad or lonely. We are all in this together. Hugs."[39] While Squeaker Tweeker uses Urban Dictionary to identify social distancing anxiety and concerns about being alone, ASMRtists and viewers work to produce a setting of shared and magnified feelings.[40] Participants imagine that this position protects individuals from COVID-19 at the same time as it materializes the pandemic in the form of videos, tactility, and feelings.

ASMRtists like Seafoam Kitten's ASMR critically consider the functions of the ASMR form. In such texts as her "ASMR - YOU ARE MY CAMERA ~ Personal Attention and Affection! Lens Touching, Button Tapping ~," feeling is connected to and understood as a means of thinking.[41] This is highlighted when Bruh notes the "metaness, and it's so tingly too."[42] For nmspy, the video "feels...meta."[43] Samurai LAN reflects that "technically we've always been the camera."[44] Bruh and nmspy suggest how critical acknowledgments of the structural components of texts still make viewers feel things. Meta-critical addresses enable individuals to be excited that mediated structures are acknowledged, to be proud that their knowledge of the constructed features of texts are recognized, and to feel important in being established as insiders and more intelligent viewers. In these cases, ASMR addresses are reflexive. Meta, as these commenters suggest, can enable forms of pleasurable thinking, facilitate visual displeasure, or both, as descriptions of scratches and gouged eyes indicate. In enacting meta positions, ASMRtists are performing a form of theory and considering the varied production processes, renderings of individuals, and forms of reading that occur with videos.

Roland Barthes, as I have previously suggested, offers a theory of reading and feeling photography. I reshape his emphasis on affective viewing and the ways cameras and photography processes render individuals, as I indicate in chapter 4, to consider ASMR and the ways it is understood to influence and reshape bodies. When Barthes feels himself "observed by the lens, everything changes."[45] He makes "another body" for himself and transforms himself "into an image." Therefore, the camera constructs and mandates certain experiences of body and self and prompts a series of sensations. Barthes's emphasis on how he feels observed and how this framework causes him to enact a series of positions is associated with the ways ASMR viewers feel the touch of producers. Being technologically viewed can constitute subjects as to-be-looked-at-ness (although the objectification of women is especially heightened and systemically structured) and how people feel. To-be-looked-at-ness is thereby correlated with to-be-touched-ness, although this may constitute emotional feelings, as well as assurances that individuals can be physically touched. Barthes outlines how he remakes himself while being photographed in a text that is focused on the ways viewers affectively experience photography and are shaped by images. In a related manner, ASMRtists consider how ASMR texts make individuals into viewers and technologies and encourage them to feel.

Touching Yourself

ASMRtists continue to touch parts of their bodies during the pandemic as methods of creating sensations and as ways of seeming to connect physically with viewers. In addition, ASMRtists produce most of their videos in indoor spaces and link the associated representations of intimate home environments to that of viewers. As I suggest earlier, some viewers have reacted with concern to such representations of contact during the pandemic. Many individuals also affectively respond to health agencies and related news reporters' identification of face touching as a disease vector. In such cases, people may experience face touching as disgusting and contaminating, which continues the cultural association of touching with disease. They are interested in not being touched and not feeling dismayed.

A CDC video about the coronavirus pandemic engages with such concerns and prompts viewers to "Learn more about the magic of handwashing."[46] The CDC thus risks portraying handwashing as a magical act that can facilitate miraculous health. Regular handwashing is a preventive practice that may be "one of the best ways to remove germs, avoid getting sick, and prevent the spread of germs to others," but it cannot inherently keep people disease or virus free.[47] Nevertheless, companies and institutions also feature hands and mandates about cleaning in pandemic opening plans. Some critics have referred to such messaging as "hygiene theater."[48] These procedures establish protocols for cleaning surfaces and related good feelings about safety, even though the major form of coronavirus spread is thought to occur through people breathing in and out droplets and aerosols that contain the virus.

Ian Carleton Schaefer and Alison E. Gabay's "Rules of the Road: Return to Work in the Time of COVID-19" emphasizes handwashing. They inform readers, "Everyone should already be washing his or her hands – all the time. This is a cardinal hygiene rule that everyone learns in elementary school."[49] The directive to "wash your hands" fits into familiar restaurant frameworks for health and picks up on people's childhood (and parenting) experiences. It is also part of colonialist legacies of evaluating cultures' intellectual and technological status through their hygiene practices.[50] Such instructions are thus aligned with, rather than an evisceration of, norms. Cleaning protocols and corporeal practices allow individuals and corporations to render their bodies, workers, and businesses as clean and safe at a time when other bodies, including people of color, continue to be rendered as threatening and deadly. For example, people with Asian lineage continue to be targeted because of the racist conflation of all individuals who appear to be Asian, the misleading identification of the coronavirus as the "Chinese virus," and the associated idea that Asian individuals are infectious.

A number of journalists use hands as methods of delimiting the virus and push hands away as schemes for trying to establish safety at a distance. For instance, Sheryl N. Hamilton's research on pandemic handshaking indicates how people are informed that their skin is tainted and "every touch is volatile, even dangerous."[51] Jenny Gross, in her *New York Times* article, makes directions on "How to Stop Touching Your Face" more normatively productive by associating such tainted touching with women's bodies and actions.[52] In a related manner, Natasha Piñon's piece "Worried About Coronavirus? Stop Touching Your Face" is illustrated with Constantin Joffe's period image of a group of women touching their faces while learning how to perform facial massage.[53] The article by Gross is illustrated with a closely cropped image of a black woman touching her eye and mouth. A red circle with a slash through it cyclically appears and warns readers not to practice such actions, but at the same time, it corrects and cancels the woman of color. Close-ups focus on the woman's skin texture and natural hair and emphasize tactility. Yet such tactility is dismissed each time the red sign appears over her face and suggests, "do not touch your face," "do not touch this face," and "do not be this person." This formulation renders black women as representations of improper actions and natural hair as too touchable and textured—a racist evaluation that is still propagated.[54] The associated directives for women to think about their physiognomic practices are related to cultural mandates for and dismissals of feminine beauty and maintenance. While there are many reasons people touch their faces, earlier conventions and contemporary news articles about the coronavirus depict this practice as feminine and vain, and nevertheless contrarily as a directive for women to fix and beautify their physiognomy and attitude.

Gross wonders, "Now that we know that it's bad to touch our faces, how do we break a habit that most of us didn't know we had?" She thereby frames this as difficult learning about negative characteristics. While individuals are encouraged not to touch their own bodies (in a way that evokes and repudiates self-pleasuring), Gross continues to identify how people "touch a lot" of contaminated surfaces. She associates touch and the environment with danger and contamination. In a revised version of stranger danger, which is also too often associated with blackness, the virus is imagined to be invisible but lurking. The initial illustration, which cancels out the black woman, also falsely correlates the out-of-control body, blackness, and "bad touching." Through this equation, women of color's active positions are

also negated. However, managed forms of touch and the processes of keeping "your hands busy," and away from your face, are conveyed through
Erin Schaff's photograph of a white man's hand holding a stress ball. His
professional and elevated positions are highlighted by including a visible
suit cuff and tie. Thus, gendered scripts about the bad body and bad touching are rendered as feminine and of color, while the controlled body and
safety are depicted as masculine and white. Such texts continue the conventions that I consider in the introduction and earlier parts of this book
by distinguishing degraded feminine embodiment and emotionality from
the elevated masculine mind and rationality.

The reporter Kalhan Rosenblatt's "Try not to touch your face. Also, try not
to think about touching your face" further correlates women, femininity, and
contagion.[55] Her article is accompanied by Jackson Gibbs's illustration of
a woman in a mask (figure A.1). The woman has a panicked look, enlarged
pupils, and studiously examines her hands. The illustration suggests that
the depicted woman has failed the mandate not to touch her face and thus
feels endangered. News agencies, as my previous comments suggest, tend
to illustrate narratives about coronavirus concerns with images of women.
This is similar to how women are represented as the drivers of fears about
dirt, as I indicate in chapter 2. In Gibbs's illustration, the woman looks

Figure A.1
Screenshot from Jackson Gibbs, "Try not to touch your face. Also, try not to think
about touching your face."

fearfully at her hands as drops of sweat bead her forehead. A series of hands dripping with various fluids circle around her and act as representations of her emotional concerns. The fingers of this series of hands are directed at her and seem to reach out to touch her without her consent. Thus, journalists' employment of hands as methods of delimitating the virus become representations of nonconsensual touching.

Not Touching and Feeling

A great deal of the psychology and medical literature on touch emphasizes the necessary and positive aspects of contact.[56] Some business and social science literature indicates that people are more likely to respond positively, including tipping greater amounts for services, when individuals socially touch them.[57] What is less frequently studied is people's negative responses to touch. A study by Jeroen Camps, Chloé Tuteleers, Jeroen Stouten, and Jill Nelissen indicates that in competitive settings, touch may lead to negative responses and outcomes.[58] While often not addressed as part of this research, scholarship on gendered violence, sexual harassment, and rape conveys instances where painful and violating touch can have long-term emotional and physical ramifications. Women's reactions to unwelcome sexual advances and touching include feeling threatened, demeaned, and shamed.[59] Some journalists address such violating and intimidating experiences and suggest that women are (or were) relieved to be working from home during the coronavirus pandemic because of uninvited touching and other forms of sexual harassment in workplaces.[60] However, other assessments indicate that sexual harassment has continued, and in some cases has even been facilitated, by telecommuting and other forms of telework.[61]

Working from home is also a problem for people caught in violent relationships. The tendency of abusers to isolate targeted individuals was elided by cultural expectations that people would remain in their homes during the first months (or year) of the pandemic and during subsequent lockdowns, and that some at-risk groups would continue this practice.[62] As Jilly Boyce Kay argues, the "injunction to 'stay the fuck at home' may work to conceal pervasive forms of gendered violence within domestic space" and incorrectly suggest that the "private, capitalist home" is a "place of safety and stability."[63] Cultural conceptions of the coronavirus have tended to escalate notions that the home is a safe place at the same time as it is the location where many people are experiencing economic precarity, depression, tight quarters, infections, and violence. Idealistic notions of home are also inaccessible to some people, including individuals who are homeless and working in other individuals' homes.

A number of journalists who mourn pandemic-related changes in the ways people physically interact have dismissed concerns about casual and uninvited contact. For instance, Jonathan Chadwick mentions #MeToo in the article "Handshakes and public hugs could go extinct in human society when the coronavirus pandemic ends, scientists warn."[64] Chadwick cites the social scientist Robert Dingwall's indication that physical distancing will not continue, although it "was already increasing as a result of #MeToo."[65] Such prognostications use the cultural losses associated with the pandemic to dismiss #MeToo and to suggest that activists' calls for respect and consent will dissipate. There is a tendency to blame women for cultural shifts in the ways people physically engage and to suggest that women enact unreasonable and uneven proscriptions against touching. For instance, a *Star-Ledger* article mentions the cultural belief that women's distinction "between 'sexual harassment' and flirting is most often based on the looks of the 'harasser.'"[66] These notions figure women as inherently heterosexual, available, cruel, and finicky. Such formulations associate gender biases and abuses with women. The associated beliefs are part of online men's groups and anti-feminists' claims that it is actually men who are culturally disenfranchised and in need of more rights.[67]

Ongoing male supremacist narratives insist that women's assertions of embodied agency and rape reporting, which supremacists swear are often fabricated, and investigations, which they insist are ordinarily biased, have resulted in a culture of cowed and endangered men. In most of these descriptions, the kinds of touching that women are supposed to have proscribed remain unspecified. What also remains unaddressed is that men's longing for intimate touch is sometimes organized around the ability to carelessly touch women (rather than physically touching men, who are often depicted as missing this contact). Society continues to associate men's acts of touching other men with gay sexualities and to identify such interactions as threatening to normative masculinities. This has influenced contemporary figurations of touch, as I suggest in earlier chapters, and appears to have informed contemporary reporting on the displacement of touch because of the coronavirus. In pandemic accounts, dismissive references to #MeToo make women's concerns about being deprived of their agency and being sexually assaulted into unreasonable anxiety about casual contact. Yet #MeToo and related activist engagements are everyday reminders that women and other disenfranchised people continue to be raped, sexually assaulted, and threatened while speaking up and living their lives.

As my comments have started to suggest, a group of journalists and other people blame women's attempts to eliminate gender-based violence for rendering contemporary proscriptions against touch and related pandemic constraints. They thereby render #MeToo and associated activist protests as unreasonable, women's claims as suspect, and expectations of bodily autonomy as too complicated. As Nickie D. Phillips and Nicholas Chagnon's research on campus sexual assault and #MeToo indicates, people invert victim-centered perspectives that would affirm notions of rape culture as a means of suggesting that heterosexual and other hegemonic relationships are threatened and "counter-hegemonic movements" are "dominating and oppressive."[68] The correlation of #MeToo with the infringement of empowered people's rights and ability to touch and connect perpetuates the cultural tradition of blaming women, feminism, and feminists for societal problems. Kristin J. Anderson, Melinda Kanner, and Nisreen Elsayegh note that feminism has been blamed for cultural shifts and catastrophes, including men's lower college enrollment numbers.[69]

In a familiar narrative, which I have started to outline, journalists and other individuals claim that women's assertions of bodily integrity have a social cost. They also establish these feminist interventions as a prehistory of the coronavirus, and thereby link women's activism to contagion and global risk and disaster. For example, Cathrine Jansson-Boyd considers the coronavirus but starts by arguing that the "decline in touch is primarily due to a fear that it may result in an accusation of inappropriate touching."[70] Carol Kinsey Goman asserts that societal "touch phobia" began with the "#MeToo movement."[71] And Mandy Oaklander creates a lineage that blames #MeToo activists when claiming, "Social hugging was largely sidelined by the Me Too movement, and smartphones took care of the rest."[72] However, such lineages are temporally confusing because these phones were mainstreamed before the use of the #MeToo hashtag in 2017. Mobile phones are also key to #MeToo activists' employment of and visibility because of social media.

Sara Ahmed's analysis of how women are discouraged from identifying problems and protesting, which I have outlined in previous parts of this book, is useful in considering the ways women are framed as being at fault for constituting such activism as #MeToo, and thereby supposedly creating problems for men. Of course, #MeToo was started and continues to be referenced because other systems to register violations and seek support and protection were deficient. Holistic addresses to #MeToo also relate the actor, producer, and activist Alyssa Milano and other recent adopters of the term to the activist work of Tarana Burke, who was not initially cited for her articulation of and earlier use of the phrase "Me Too." In all of these instances, #MeToo is a means of highlighting cultural and gendered wrongs by a kind of symbolic raising of hands in recognition of and solidarity with other women. The communication scholar Rosemary Clark-Parsons describes #MeToo, and hashtag feminism more broadly, as the methods that activists employ to "make the personal political by making it visible, bridging the individual with the collective and illustrating the systemic nature of social injustice."[73] Feminist academics, including Carrie A. Rentschler, have also expressed concern about the ways #MeToo and related politics risk advocating for the continuance of carceral practices rather than interventions that emerge from and are resolved within specific communities.[74]

Many individuals identify the #MeToo hashtag and related actions as repressing hand-to-body and body-to-body contact. However, online #MeToo posts are often accompanied by images of groups of raised hands, which envision interpersonal and community activism and connections. Ahmed indicates how women's arms and hands have been used as political symbols that encourage women to collaborate and refuse the ways society has controlled their labor. Women's hands are too often compelled into serving, or are even possessed by, other individuals, who are often men. Ahmed rightly associates raised arms with processes of pointing to problems and asserting the rights of otherwise oppressed individuals. In these instances, women indicate the ongoing problems of sexual assault and harassment and are made into the problem. Women's protesting hands are conceptually pushed down so that other individuals can continue to experience unconstrained touch, including the right to touch women without invitation. Within this oppressive framework, women's interventions into being constituted as to-be-touched-ness are understood as the evisceration of tactile and comfortable connections. Touching is constituted as feminine and correlated with women, and women are blamed for withholding or destroying it when advocating for more ethical and respectful engagements.

Journalists' indications that #MeToo has severe costs, and the associated calls for moderation, do not take into account the particular ways women's bodies are culturally framed and delimited. This includes the ways women are figured as endangered *and* as needing to be available for physical contact. As Carey-Ann Morrison argues, "Women's bodies, in particular, are often deemed spatially open and available," which I would argue includes being understood as to-be-touched-ness.[75] Thus, "women sometimes experience an erosion of their personal boundaries and loss of corporeal freedom in sexual relationships. Domestic violence and rape are prime examples of the ways in which sexual relationships seem to confer ownership rights over the body of another," which is ordinarily identified as a woman's body.

Unwelcome touching can afford power to the toucher, including instances where men touch women without invitation, and act as an indicator that women need protection, are endangered, and are culturally constituted as being at risk of rape. As Ann J. Cahill argues, threats of rape produce the "feminine body."[76] Women are culturally constituted and encouraged to understand their flesh as "inherently weak," "breakable," and "violable." Kari Stefansen indicates that girls' persistent experiences with microtransgressions shrink their claims to spaces and work as reminders that their "right to bodily integrity is less protected than it is for boys."[77] Such renderings of women's bodies may also warn women that they need to pay particular attention to a broader group of threats rather than questioning specific perpetrators and trying to facilitate safe and viable lives for all subjects. These dismissals of women and girls suggest that their expectations are unreasonable and it is unlikely that their boundaries and bodies will be respected and governmentally and socially protected.

Conclusion: Socially Distancing as a Method of Not Feeling

Journalists emphasize and try to suspend the relationship between bodies and screens because of concerns about coronavirus infection. For instance, Jeff Link worries about the use of digital device screens in "Will We Ever Want to Use Touchscreens Again?: The future is touchless."[78] Yet the question is asked and answered in the title. The concept of an outlook without touch revisions the physical hand-to-object contact that is key to most individuals' digital device use. This framework also negates people's more overarching physical contact and touching at depth with other individuals, animate things, and objects. In such texts, crisis narratives replace past conceptions of the desirable futurity of digital communication. Such texts risk conflating the loss of tactile culture with the human toll of the pandemic and the individuals and identities that are situated as dispensable and predisposed to death, including the elderly and people of color. Related crisis frameworks claim that the associated social and infrastructural failures, including the unavailability of healthcare, are new and unique to the pandemic, rather than considering how many governments have underinvested in medical research, hospital systems, and supply chains and thus are underinvested in maintaining people's lives and health.

Link is prompted to consider cultural change when he goes into a 7-Eleven store and encounters a notice not to touch the Slurpee machine because of the risk of viral contamination. Touch is thereby associated with consumer experiences and neoliberal directives for personal responsibility rather than state support, including the displacement of medical advice about managing the pandemic and other viral hazards. The store notification encourages him to worry over possible infections from such objects and surfaces as the "touchscreen payment reader" and the "plastic surface of the smartphone case" that he touches numerous times every day. Link animates objects and foregrounds the fragility of bodies when wondering "if people would recall this time as the death knell of public touchscreens." Thus, screens are depicted as potentially dying because their cultural position is animate and alive. Such commentary also enlivens technologies by refuting the "death" of cultural norms and practices. For instance, Christian Hetrick reports, "No, coronavirus won't kill the touchscreen."[79] While Hetrick animates the screen by asserting that the pandemic will not kill it, the association of death and touch is still heightened. These narratives risk supplanting people's deaths and loss of livelihood, which have been exacerbated by state disinterest in and misinformation about best health practices, with the more palatable loss of touchscreen contact.

In the journalists' accounts that I study, physically touching and emotionally feeling are connected to good pasts that have been lost. Not touching is viscerally experienced and conceptualized in relation to physical contact. Lisa Bonos chronicles, "Some describe the lack of touch as its own sensory experience: A dull ache. Skin that hurts. A hole in the pit of the stomach."[80] Erika Hughes indicates, "There's almost something painful about not crossing that line" and touching people.[81] These texts correlate the dearth of touch with emotion and suggest that we should theorize how physically distancing and emotionally feeling are correlated. For instance, not touching produces a form of the ephemeral and experiential punctum that Barthes theorizes when viewing but not inherently touching photographs. Accounts of pulls and aches also point to ways of rethinking touching at depth so that it accounts for people's longing for and the absence of bodily contact.

People's pandemic narratives depict close contact as missed and emotionally uncomfortable. Marina Koren describes "The New Cringeworthy," where the "sight of two people shaking hands," "Someone touching their uncovered face," and a "group of people hanging out less than six feet apart" render "sudden, visceral reactions—of discomfort or disgust, fear or indignation—whether they're occurring on-screen or in real life."[82] In such instances, even the idea of physically touching is hinged to negative emotional feelings. Koren chronicles the kinds of "Eew! Factor" and "excited disgust" that Muriel Dimen outlines and that I consider in relation to dirty screens in chapter 2.[83] As Koren notes, such visceral affects move the body and are part of people's desire to push away and out other subjects and things. Julia Kristeva studies similar forms of gut-wrenching and annihilating responses to the thin skins of foodstuffs and fluids that are culturally framed as dirty.[84] Thus, ideas about and representations of hands in these accounts are repulsive, but so are the fingernails that are associated with the women who employ touchscreens. For individuals receiving new phones, as I indicate in chapter 2, even the appearance of a hair under glass (and thereby mediated) is often cause for alarm and deemed to be viscerally disgusting. Given that the experiences of dirt and repulsion that I focus on earlier in this book often trigger feelings of repulsion, such cringing is not new. It is part of a series of processes that link emotionally feeling to attempts to physically distance and not touch.

The texts that I outline in this book engage with and through digital technologies and screens, and thus render touching at depth and touching that is ordinarily at some distance from other bodies. Shanley Pierce reports on this position when proposing that "virtual alternatives can help alleviate the effects of touch starvation."[85] Yet journalists are interested in reprieving ideal experiences of touching that are not bounded by proscriptions or acknowledgments of individuals' personal limits. While the narratives that I analyze are common, there is no developed or critical literature on how physically distancing and emotionally feeling are correlated. Narratives about pulls and aches suggest a way of rethinking touching at depth so that it accounts for people's longings and figurations of bodily contact. Barthes's theorization of punctum provides methods of considering forms of physically distancing that amplify embodied feelings.

Tahhan and Barthes offer different ways of considering how to-be-touched-ness functions in people's lives. Laura Mulvey and related feminist scholars have suggested how films employ varied devices to script women as to-be-looked-at-ness and objects of male contemplation and control.[86] As the texts that I mention in this book suggest, to-be-looked-at-ness is often directed at women and constitutes young, normatively sized, and light-skinned women as of particular interest. Women of color are figured as both hypervisible and as unnoticed. The cultural formation of normative women as to-be-looked-at-ness contributes to and is associated with constituting the same subjects as to-be-touched-ness. There are cultural amplifications of this to-be-touched-ness in the ways people of color and other oppressed individuals are coerced into being the subjects of medical and other examinations.

Individuals, sites, and companies tend to intensify cultural conceptions of to-be-touched-ness when they correlate touchscreens and hands. This includes the ways women are figured as sites of touch and system failure when they refuse to shape their fingernails in the manner prescribed and to become another body. In such instances, women raise their hands as a means of refusal and self-identification. Thus, the raised hand and employment of #MeToo to identify women's experiences with nonconsensual contact are methods of action and theories that propose other ways of reading and enacting relationality. Throughout this book, I offer a set of critical theories of touching and to-be-touched-ness as a means of encouraging further considerations of how people engage with each other, as well as with and through touchscreen devices. This includes foregrounding how the coronavirus has been used as a scheme to direct women to accept their position as to-be-touched-ness. The pandemic causes people to move back from and mourn to-be-touched-ness. These gestures, which tend to disenfranchise already oppressed subjects, can be foregrounded by further identifying how particular subjects are scripted to feel and the bodies that are supposed to matter and be expendable.

While writing this book and trying to engage what Ahmed identifies as the challenges and joys of a feminist life, I have raised my arm along with and through feminist activism and scholarship and moved it as a means of pointing to the narratives, bodies, and feelings that continue to be identified as unimportant.[87] For the manufacturers, designers, and individuals who develop and employ digital devices, writing is touching and downplayed in favor of hands that appear to move through and mesh with devices. I am reminded of a colleague's repressive query about "what happened," because I "used to do such important writing." It is at such junctures between what others identify as "important," and thus scripts that reproduce their values and identities, and research and readings that they deem uncomfortable and never important that my feminist research and feelings struggle between attempts at substantiation and accepting their identification and being a problem. In the case of scholarship and Internet practices, I hope that I have demonstrated that critical interventions into what has been rendered as "unimportant" and as "problems" are worth pursuing.

Notes

Preface and Acknowledgments

1. Caleb Kelly identifies how media "technologies (computers, iPhones, tablets) are often imagined to be pristine and clean." He also describes a "sense of guilt" when placing "fingers onto it (leaving the first fingerprints of many to come)." Caleb Kelly, "Dirt(y) Media: Dirt in Ecological Media Art Practices," *European Journal of Cultural Studies* (Online First, 2021): 1–16.

2. Tim Cook, "Apple Event: 14 September 2021," Apple, 15 September 2021, https://www.apple.com/apple-events/september-2021/; For my methods of quoting online materials and formatting Internet citations, see my first endnote in the introduction.

3. Apple, Twitter, 9 September 2021, https://twitter.com/Apple

4. Conrad Bakker, "Untitled Project: Smartphone [Cracked Screen] [#2]," Flickr, 13 January 2020, 19 September 2021, https://www.flickr.com/photos/untitledprojects/49383151177/. The quoted text is from Conrad Bakker, "Untitled Project: Smartphone [#CRACKEDSCREEN]," Tumblr, 2019–2020, 15 September 2021, https://conradbakker.tumblr.com/post/190284659851/untitled-project-smartphone-crackedscreen. Bakker titles his illustrated work in a slightly different manner on Flickr and Tumblr, which is quite common with online identification.

Introduction

1. Cox, "News Releases | Cox Communications," 14 January 2020, https://newsroom.cox.com/news-releases. Many authors' Internet texts include typographical errors and unconventional forms of spelling, uppercase and lowercase typefaces, punctuation, and spacing. I have retained these formatting features in quotations and Internet references, without such qualifications as "intentionally so written" or "sic." I have left the titles of Internet articles and sites as they are represented online, but I have reformatted newspaper citations. In the references, the date listed before the URL is the "publication" date, or the last time the site was viewed in the indicated

format. When two dates are included, the first date points to when the current configuration of the site was initially available, and the second date is the latest access date. Some referenced sites are no longer available, so I used the Internet Archive's Wayback Machine to consult a version of these texts. The Wayback Machine may also offer content that was available when this book was researched but is no longer accessible. Internet Archive, "Internet Archive: Wayback Machine," 21 August 2021, https://archive.org/web/web.php

2. Apple, "iPhoneX," 17 January 2017, https://www.apple.com/iphone-x/

3. Apple, "iPhone 12 Pro and iPhone 12 pro Max Key Features," 21 June 2021, https://www.apple.com/iphone-12-pro/key-features/

4. Eve Kosofsky Sedgwick, *Touching Feeling: Affect, Pedagogy, Performativity* (Durham, NC: Duke University Press, 2003), 17.

5. M. McCormack, "The Declining Significance of Homohysteria for Male Students in Three Sixth Forms in the South of England," *British Educational Research Journal* 37, no. 2 (April 2011): 37–53.

6. Els Rommes, Ellen van Oost, and Nelly Oudshoorn, "Gender in the Design of the Digital City of Amsterdam," *Information, Communication, and Society* 2, no. 4 (1999): 476–95; Nelly Oudshoorn, Els Rommes, and Marcelle Stienstra, "Configuring the User as Everybody: Gender and Design Cultures in Information and Communication Technologies," *Science, Technology, and Human Values* 29, no. 1 (Winter 2004): 30–63.

7. Apple, "iPad4 16Gb Wi-Fi black," 21 January 2017, http://solax.hk/tablets.html

8. Apple has also asserted that "technology is at its very best when it is invisible." TouchGameplay, "Official Apple (New) iPad Trailer," YouTube, 7 March 2012, 27 June 2021, https://www.youtube.com/watch?v=RQieoqCLWDo

9. Lisa Gitelman and Geoffrey B. Pingree, "What's New About New Media?" in *New Media, 1740–1915*, ed. Lisa Gitelman and Geoffrey B. Pingree (Cambridge, MA: MIT Press, 2003), xiii.

10. Laura Mulvey, *Visual and Other Pleasures* (Bloomington: Indiana University Press, 1989).

11. Lev Grossman, "Invention of the Year: The iPhone," *Time*, 1 November 2007, 5 January 2018, http://content.time.com/time/specials/2007/article/0,28804,1677329_1678542,00.html

12. Mark Paterson, "Introduction: Remediating Touch," *Senses and Society* 4, no. 2 (2009): 129–40.

13. Maurice Merleau-Ponty's mid-twentieth-century phenomenological analysis of touching has influenced contemporary scholarship. Merleau-Ponty argues that when hands or other body parts are in contact, that it is impossible to differentiate

between what is touching and what is being touched. Maurice Merleau-Ponty, *Signs*, trans. Richard C. McCleary (Evanston, IL: Northwestern University Press, 1964).

14. Diana Adis Tahhan, "Touching at Depth: The Potential of Feeling and Connection," *Emotion, Space and Society* 7 (2013): 45.

15. Tahhan, "Touching at Depth," 46.

16. Tahhan, "Touching at Depth," 49.

17. Tahhan, "Touching at Depth," 50.

18. Laura U. Marks, "Video Haptics and Erotics," *Screen* 39, no. 4 (1998): 331–47.

19. Laura U. Marks, *Touch: Sensuous Theory and Multisensory Media* (Minneapolis: University of Minnesota Press, 2002).

20. Vivian Sobchack, "What My Fingers Knew: The Cinesthetic Subject, or Vision in the Flesh," *Senses of Cinema* 5 (2000): http://sensesofcinema.com/2000/conference -special-effects-special-affects/fingers/. The italic emphasis appears in Sobchack's text.

21. Kevin E. McHugh, "Touch at a Distance: Toward a Phenomenology of Film," *GeoJournal* 80 (2015): 840.

22. John Cromby and Martin E. H. Willis, "Affect—or Feeling (After Leys)," *Theory and Psychology* 26, no. 4 (2016): 476–95.

23. Ruth Leys, "The Turn to Affect: A Critique," *Critical Inquiry* 37 (Spring 2011): 434–72.

24. Respondents to Ley, including Charles Altieri, have reasserted that affect is not utterly distinct from intentionality and consciousness, and thereby they highlighted the different ways affect is defined and employed. Charles Altieri, "Affect, Intentionality, and Cognition: A Response to Ruth Leys," *Critical Inquiry* 38 (Summer 2012): 878–81.

25. Susanna Paasonen, Ken Hillis, and Michael Petit, "Introduction: Networks of Transmission: Intensity, Sensation, Value," in *Networked Affect*, ed. Ken Hillis, Susanna Paasonen, and Michael Petit (Cambridge, MA: MIT Press, 2015), 1–25. Their work is informed by Gregory J. Seigworth and Melissa Gregg's description of affect as what "arises in the midst of in-between-ness," the intensities that "pass body to body (human, non-human, part-body and otherwise)," and the "resonances that circulate about, between, and sometimes stick to bodies and worlds." Gregory J. Seigworth and Melissa Gregg, "An Inventory of Shimmers," in *The Affect Theory Reader*, ed. Melissa Gregg and Gregory J. Seigworth (Durham, NC: Duke University Press, 2010), 1.

26. Paasonen, Hillis, and Petit, "Introduction," 1.

27. Susanna Paasonen, "A Midsummer's Bonfire: Affective Intensities of Online Debate," in *Networked Affect*, ed. Ken Hillis, Susanna Paasonen, and Michael Petit (Cambridge, MA: MIT Press, 2015), 27–42.

28. Sara Ahmed, "Affective Economies," *Social Text* 22, no. 2 (79) (Summer 2004): 119.

29. Ahmed, "Affective Economies," 128. The italic emphasis appears in Ahmed's text.

30. Apple, "iPhoneX," 17 January 2017, https://www.apple.com/iphone-x/

31. Heidi Rae Cooley, "It's All About the Fit: The Hand, the Mobile Screenic Device and Tactile Vision," *Journal of Visual Culture* 3, no. 2 (2004): 133–55.

32. Donna Haraway, *Simians, Cyborgs, and Women: The Reinvention of Nature* (New York: Routledge, 1991).

33. David Parisi's historical study of touchscreens relates engagements with such devices to midcentury attempts to facilitate individuals' engagements in "distant environments using devices that mimicked the motions and actions of the human hands." David Parisi, *Archaeologies of Touch: Interfacing with Haptics from Electricity to Computing* (Minneapolis: University of Minnesota Press, 2018), 214.

34. Jack Bratich, "The Digital Touch: Craft-work as Immaterial Labour and Ontological Accumulation," *Ephemera: Theory and Politics in Organization* 10, nos. 3–4 (2010): 303.

35. Shaun Moores, "Digital Orientations: 'Ways of the Hand' and Practical Knowing in Media Uses," *Mobile Media and Communication* 2, no. 2 (2014): 204.

36. Sarah Pink, Jolynna Sinanan, Larissa Hjorth, and Heather Horst, "Tactile Digital Ethnography: Researching Mobile Media Through the Hand," *Mobile Media and Communication* 4, no. 2 (2016): 237–51.

37. Michele White, *Producing Masculinity: The Internet, Gender, and Sexuality* (New York: Routledge, 2019).

38. Peter J. Capuano, *Changing Hands: Industry, Evolution, and the Reconfiguration of the Victorian Body* (Ann Arbor: University of Michigan Press, 2015), 130; Jacques Derrida, *On Touching—Jean-Luc Nancy*, trans. Christine Irizarry (Stanford, CA: Stanford University Press, 2005), 185. The italic emphasis appears in Derrida's text.

39. J. Hillis Miller, "Derrida Enisled," *Critical Inquiry* 33, no. 2 (Winter 2007): 248–76; Martin Heidegger, *What Is Called Thinking?* trans. J. Glenn Gray (New York: HarperCollins Publishers, 1976), 16.

40. Capuano, *Changing Hands*, 238–39.

41. Jean-Pierre V. M. Hérubel, "The Darker Side of Light: Heidegger and Nazism: A Bibliographic Essay," *Shofar* 10, no. 1 (Fall 1991): 85–105.

42. Tom Tyler, "The Rule of Thumb," *JAC* 30, nos. 3–4 (2010): 435–56.

43. Tyler, "Rule of Thumb," 451.

44. Michele White, *The Body and the Screen: Theories of Internet Spectatorship* (Cambridge, MA: MIT Press, 2006).

45. Apple, "iPad4 16Gb Wi-Fi black," 21 January 2017, http://solax.hk/tablets.html

46. Microsoft, "Mouse and Pointers," 13 September 2018, https://docs.microsoft.com/en-us/windows/desktop/uxguide/inter-mouse

47. Apple Developer, "Mouse and Trackpads," Human Interface Guidelines, 25 January 2019, https://developer.apple.com/design/human-interface-guidelines/macos/user-interaction/mouse-and-trackpad/

48. Microsoft, "Windows 7 Mouse and Pointers," 7 February 2022, 12 March 2022, https://docs.microsoft.com/en-us/windows/desktop/uxguide/inter-mouse

49. Susan Kare, "About," 11 September 2018, http://kare.com/about/; Susan Kare, "Apple," 11 September 2018, http://kare.com/apple-icons/Kare

50. Susan Kare, "Interview with Susan Kare," Making the Macintosh: Technology and Culture in Silicon Valley, 20 February 2001, 11 September 2018, https://web.stanford.edu/dept/SUL/sites/mac/primary/interviews/kare/trans.html

51. Nicholas Sammond, *Birth of an Industry: Blackface Minstrelsy and the Rise of American Animation* (Durham, NC: Duke University Press, 2015), 2–3.

52. Kit Grose, "Who created the Mac Mickey pointer cursor?" User Experience Stack Exchange, 15 September 2016, 30 January 2021, https://ux.stackexchange.com/questions/52503/who-created-the-mac-mickey-pointer-cursor

53. The White Files, "Resource Types," 20 July 2012, http://www.whitefiles.org/b1_s/1_free_guides/fg3mo/pgs/t02_rslst.htm; a.jfred, "Do you think the mouse pointer hand in Mac OS relates to Steve Jobs and Disney?" MacRumors, 8 August 2011, 30 July 2012, http://forums.macrumors.com/showthread.php?t=1208582

54. Mulvey, *Visual and Other Pleasures*.

55. Mulvey, *Visual and Other Pleasures*, 25.

56. Richard Dyer, "The Colour of Virtue: Lillian Gish, Whiteness, and Femininity," in *Women and Film: A Sight and Sound Reader*, ed. Pam Cook and Philip Dodd (Philadelphia: Temple University Press, 1993), 1–9; Patrick Keating, *Hollywood Lighting from the Silent Era to Film Noir* (New York: Columbia University Press, 2009).

57. Elizabeth S. Leet, "Objectification, Empowerment, and the Male Gaze in the Lanval Corpus," *Historical Reflections* 42, no. 1 (Spring 2016): 81.

58. Apple, "Apple – Portraits of Her," Facebook, 29 September 2017, 27 January 2021, https://www.facebook.com/watch/?v=10214325484908461; SouL Gaming, "iPhone 8 Plus – Portraits of Her – Apple," YouTube, 6 October 2017, 21 January 2021, https://www.youtube.com/watch?v=nogm_UzDZSA

59. AdStasher, "Apple's 'Portraits of Her' Commercial Shows Off New Portrait Lighting Genius of the iPhone 8 Plus," 8 October 2017, 8 July 2021, https://www.adstasher.com/2017/10/apples-portraits-of-her-commercial.html?m=0

60. Apple, "iPhone 12 and iPhone 12 mini Key Features," 6 July 2021, https://www.apple.com/iphone-12/key-features/

61. Barbara Johnson, "Teaching Deconstructively," in *Writing and Reading Differently*, ed. George Douglas Atkins, Michael L. Johnson, and Nancy R. Comley (Lawrence: University of Kansas Press, 1986), 140.

62. N. Katherine Hayles, *How We Think: Digital Media and Contemporary Technogenesis* (Chicago: University of Chicago Press, 2012), 2.

63. Sara Ahmed, *Living a Feminist Life* (Durham, NC: Duke University Press, 2017).

64. Matthew Kirschenbaum, *Mechanisms: New Media and the Forensic Imagination* (Cambridge, MA: MIT Press, 2008).

65. Hayles, *How We Think*, 58.

66. Nicole Shukin, "The Hidden Labour of Reading Pleasure," *English Studies in Canada* 33, nos. 1–2 (March–June 2007): 23–27.

67. N. Katherine Hayles, "Hyper and Deep Attention: The Generational Divide in Cognitive Modes," *Profession* (2007): 187.

68. Hyper reading evokes speed and the forms of hyper attention that are engaged when cycling between different windows, content, and forms of media texts. Machine reading is a kind of distant or not reading and occurs when texts are "read" automatically and without human supervision. However, machine reading is still influenced by human perceptions and worldviews since programmers and other individuals determine the associated algorithmic schematics and coding.

69. Jonathan Culler, *Literary Theory: A Very Short Introduction* (New York: Oxford University Press, 1997).

70. Nanna Verhoeff, "Theoretical Consoles: Concepts for Gadget Analysis," *Journal of Visual Culture* 8, no. 3 (2009): 279–98.

71. Jordan Alexander Stein, *Avidly Reads Theory* (New York: New York University Press, 2019).

72. Terry Eagleton, *How to Read a Poem* (Malden, MA: Blackwell Publishing, 2007), 2.

73. Michelle Marzullo, Jasmine Rault, and T. L. Cowan, "Can I Study You? Cross-Disciplinary Studies in Queer Internet Studies," *First Monday* 23, no. 7, 2 July 2018, 7 August 2020, http://www.firstmonday.dk/ojs/index.php/fm/article/view/9263/7465

74. Jasmine Rault, as cited in Michelle Marzullo, Jasmine Rault, and T. L. Cowan, "Can I Study You? Cross-Disciplinary Studies in Queer Internet Studies," *First Monday* 23, no. 7, 2 July 2018, 7 August 2020, http://www.firstmonday.dk/ojs/index.php/fm/article/view/9263/7465

75. Annette Markham, Elizabeth Buchanan, and AoIR Ethics Working Committee, "Ethical Decision-Making and Internet Research: Version 2.0," Association of Internet Researchers, 2012, 8 July 2018, https://aoir.org/ethics/; aline Shakti franzke, Anja Bechmann, Michael Zimmer, Charles Ess, and the Association of Internet Researchers, "Internet Research: Ethical Guidelines 3.0," 6 October 2019, 3 February 2021, https://aoir.org/reports/ethics3.pdf

76. Ann J. Cahill, "Foucault, Rape, and the Construction of the Feminine Body," *Hypatia* 15, no. 1 (2000): 43–63.

77. Michele White, "GIFs from Feminists: Visual Pleasure, Danger, and Anger on the Jezebel Website," *Feminist Formations* 30, no. 2 (Summer 2018): 202–30.

78. Michele White, "Representations or People?" *Ethics and Information Technology* 4, no. 3 (2002): 249–66.

79. Elaine Ginsberg and Sara Lennox, "Antifeminism in Scholarship and Publishing," in *Antifeminism in the Academy*, ed. Veve Clark, Shirley Nelson Garner, Margaret Higonnet, and Ketu Katrak (New York: Routledge, 2014), 169–99.

80. Jevin D. West, Jennifer Jacquet, Molly M. King, Shelley J. Correll, and Carl T. Bergstrom, "The Role of Gender in Scholarly Authorship," *PLoS One* 8, no. 7 (2013): 1–6; Marieke Van den Brink and Yvonne Benschop, "Gender Practices in the Construction of Academic Excellence: Sheep with Five Legs," *Organization* 19, no. 4 (2012): 507–24.

81. Michelle Quinn, "Latest Technology Is at Your Fingertips," *Los Angeles Times*, 4 February 2007, 14 May 2018, http://www.latimes.com/la-fi-iphone4-2008feb04-story.html

82. Madeleine Akrich, "The De-scription of Technical Objects," in *Shaping Technology/Building Society: Studies in Sociotechnical Change*, ed. Wiebe E. Bijker and John Law (Cambridge, MA: MIT Press, 1994), 205–24; Oudshoorn, Rommes, and Stienstra, "Configuring the User."

83. Nelly Oudshoorn and Trevor Pinch, *How Users Matter: The Co-construction of Users and Technologies* (Cambridge, MA: MIT Press, 2003).

84. Ahmed, *Living a Feminist Life*.

85. Sarah Banet-Weiser, *Empowered: Popular Feminism and Popular Misogyny* (Durham, NC: Duke University Press, 2018); Mulvey, *Visual and Other Pleasures*; White, *Body and the Screen*.

86. Sara Ahmed, "The Skin of the Community: Affect and Boundary Formation," in *Revolt, Affect, Collectivity: The Unstable Boundaries of Kristeva's Polis*, ed. Tina Chanter and Ewa Plonowska Ziarek (Albany: State University of New York Press, 2012), 95–111; Didier Anzieu, *The Skin-ego*, trans. Naomi Segal (London: Karnac Books, 2016);

Nicolette Bragg, "'Beside Myself': Touch, Maternity and the Question of Embodiment," *Feminist Theory* 21, no. 2 (2020): 141–55; Naomi Segal, "'A Petty Form of Suffering': A Brief Cultural Study of Itching," *Body and Society* 24, nos. 1–2 (2018): 88–102.

87. Roland Barthes, *Camera Lucida: Reflections on Photography*, trans. Richard Howard (New York: Farrar, Strauss, and Giroux, 1981); Steve Woolgar, "Configuring the User: The Case of Usability Trials," in *A Sociology of Monsters: Essays on Power, Technology and Domination*, ed. John Law (London: Routledge, 1991), 58–99.

88. Craig Richard, *Brain Tingles: The Secret to Triggering Autonomous Sensory Meridian Response for Improved Sleep, Stress Relief, and Head-to-Toe Euphoria* (New York: Simon and Schuster, 2018).

89. Michele Hilmes, "The Television Apparatus: Direct Address," *Journal of Film and Video* 37, no. 4 (Fall 1985): 27–36.

90. World Health Organization, "Naming the coronavirus disease (COVID-19) and the virus that causes it," 9 November 2020, https://www.who.int/emergencies /diseases/novel-coronavirus-2019/technical-guidance/naming-the-coronavirus -disease-(covid-2019)-and-the-virus-that-causes-it

91. Tahhan, "Touching at Depth"; Barthes, *Camera Lucida*.

92. Mulvey, *Visual and Other Pleasures*.

Chapter 1

1. MacRumors, "MacRumors: Apple Mac iPhone Rumors and News," 6 December 2020, https://www.macrumors.com/

2. katie ta achoo, "Are Long Fingernails Compatible with iPhone?" MacRumors, 9 July 2007, 29 December 2018, https://forums.macrumors.com/threads/are-long-finger nails-compatible-with-iphone.327743/

3. HotdogGiambi, "Are Long Fingernails Compatible with iPhone?" MacRumors, 11 July 2007, 15 May 2018, https://forums.macrumors.com/threads/are-long-fingernails -compatible-with-iphone.327743/

4. MrSmith, "Are Long Fingernails Compatible with iPhone?" MacRumors, 11 July 2007, 15 May 2018, https://forums.macrumors.com/threads/are-long-fingernails-com patible-with-iphone.327743/

5. Madeleine Akrich, "The De-scription of Technical Objects," in *Shaping Technology/Building Society: Studies in Sociotechnical Change*, ed. Wiebe E. Bijker and John Law (Cambridge, MA: MIT Press, 1994), 205–24; Nelly Oudshoorn, Els Rommes, and Marcelle Stienstra, "Configuring the User as Everybody: Gender and Design Cultures in Information and Communication Technologies," *Science, Technology, and Human Values* 29, no. 1 (Winter 2004): 30–63.

6. Nelly Oudshoorn and Trevor Pinch, *How Users Matter: The Co-construction of Users and Technologies* (Cambridge, MA: MIT Press, 2003).

7. Michelle Quinn, "Latest Technology Is at Your Fingertips," *Los Angeles Times*, 4 February 2007, 14 May 2018, http://www.latimes.com/la-fi-iphone4-2008feb04 -story.html; Michelle Quinn, "The iPhone Fingernail Problem," *Los Angeles Times*, 12 June 2008, 14 May 2018, http://latimesblogs.latimes.com/technology/2008/06 /the-fingernail.html

8. Sara Ahmed, *Living a Feminist Life* (Durham, NC: Duke University Press, 2017).

9. Sarah Sobieraj and Jeffrey M. Berry, "From Incivility to Outrage: Political Discourse in Blogs, Talk Radio, and Cable News," *Political Communication* 28, no. 1 (2011): 19–41.

10. Jack Bratich, "The Digital Touch: Craft-work as Immaterial Labour and Ontological Accumulation," *Ephemera: Theory and Politics in Organization* 10, nos. 3–4 (2010): 303–18.

11. Gerard Goggin, "Disability and Haptic Mobile Media," *New Media and Society* 19, no. 10 (2017): 1563–580.

12. Oudshoorn, Rommes, and Stienstra, "Configuring the User."

13. Steve Woolgar, "Configuring the User: The Case of Usability Trials," in *A Sociology of Monsters: Essays on Power, Technology and Domination*, ed. John Law (London: Routledge, 1991), 58–99.

14. Oudshoorn, Rommes, and Stienstra, "Configuring the User," 32.

15. Silvan S. Tomkins, "Script Theory: Differential Magnification of Affects," in *Nebraska Symposium on Motivation*, ed. H. E. Howe and R. A. Dienstbier (Lincoln: University of Nebraska Press, 1978), 201–36.

16. Akrich, "De-scription of Technical Objects," 208.

17. Majken Kirkegaard Rasmussen and Marianne Graves Petersen, "Re-scripting Interactive Artefacts with Feminine Values," in *Proceedings of the 2011 Conference on Designing Pleasurable Products and Interfaces*, Milan: ACM, 2011, 1–8.

18. Els Rommes, Ellen van Oost, and Nelly Oudshoorn, "Gender in the Design of the Digital City of Amsterdam," *Information, Communication, and Society* 2, no. 4 (1999): 479.

19. Els Rommes, "Creating Places for Women on the Internet: The Design of a 'Women's Square' in a Digital City," *European Journal of Women's Studies* 9, no. 4 (2002): 413.

20. Brothers Grimm, "The Willful Child," in *The Complete Grimm's Fairy Tales, trans. Margaret Hunt* (New York: Pantheon Books, 1944), no. 117.

21. Ahmed, *Living a Feminist Life*, 79.

22. Ahmed, *Living a Feminist Life*, 84.

23. Ahmed, *Living a Feminist Life*, 85.

24. Ahmed, *Living a Feminist Life*, 38.

25. Ahmed, *Living a Feminist Life*, 39.

26. Jennifer Schumann, Sandrine Zufferey, and Steve Oswald, "What Makes a Straw Man Acceptable? Three Experiments Assessing Linguistic Factors," *Journal of Pragmatics* 141 (2019): 1.

27. Robert Talisse and Scott F. Aikin, "Two Forms of the Straw Man," *Argumentation* (2006): 345–52.

28. Barbara Johnson, "Teaching Deconstructively," in *Writing and Reading Differently*, ed. George Douglas Atkins, Michael L. Johnson, and Nancy R. Comley (Lawrence: University of Kansas Press, 1986), 140–48.

29. Ellen Goodman, "Straw Feminist Declares Open Season on Men," *Baltimore Sun*, 26 January 1994, 30 August 2020, https://www.baltimoresun.com/news/bs-xpm-1994 -01-26-1994026037-story.html

30. Monica Dux and Zora Simic, *The Great Feminist Denial* (Melbourne: Melbourne University Press, 2008), 6.

31. Sobieraj and Berry, "From Incivility to Outrage."

32. Sobieraj and Berry, "From Incivility to Outrage," 39–41.

33. CBS Sunday Morning, "Happy 10th birthday, iPhone," YouTube, 25 June 2017, 14 March 2022, https://www.youtube.com/watch?v=aHjFKqzUpII&t=1s

34. David Pogue, "The iPhone Matches Most of Its Hype," *New York Times*, 27 June 2007, 6 January 2018, https://www.nytimes.com/2007/06/27/technology/circuits/27pogue .html?ex=1340596800&en=d00bbea4b9e0ece6&ei=5124&partner=permalink &exprod=permalink

35. Walter S. Mossberg and Katherine Boehret, "Testing Out the iPhone," *Wall Street Journal*, 27 June 2007, 5 August 2019, https://www.wsj.com/articles /SB118289311361649057

36. Walter S. Mossberg, as cited in CBS News, "It had us at 'Hello': The iPhone turns 10," 25 June 2017, 5 January 2018, https://www.cbsnews.com/news/the-iphone -turns-10/

37. Steve Jobs, as cited in Peter Cohen, "Macworld Expo Keynote Live Update: Introducing the iPhone," Macworld, 9 January 2007, 5 January 2018, https://www .macworld.com/article/1054764/macworld-expo/liveupdate.html

38. Apple, "Apple Pencil," 12 January 2018, https://www.apple.com/apple-pencil/

39. Michelle Quinn, "Latest Technology Is at Your Fingertips," *Los Angeles Times*, 4 February 2007, 14 May 2018, http://www.latimes.com/la-fi-iphone4-2008feb04-story .html

40. Natalie Kerris, as cited in Michelle Quinn, "Latest Technology Is at Your Fingertips," *Los Angeles Times*, 4 February 2007, 14 May 2018, http://www.latimes.com/la -fi-iphone4-2008feb04-story.html

41. David Pogue, "The iPhone Matches Most of Its Hype," *New York Times*, 27 June 2007, 4 November 2018, https://www.nytimes.com/2007/06/27/technology/circuits /27pogue.html?pagewanted=all&_r=0

42. Michelle Quinn, "Finally, Steve Jobs Unveils iPhone 2.0 and iPhone 3G," *Los Angeles Times*, 9 June 2008, 28 July 2021, https://latimesblogs.latimes.com /technology/2008/06/finally-steve-j.html

43. Erica, "Finally, Steve Jobs Unveils iPhone 2.0 and iPhone 3G," *Los Angeles Times*, 9 June 2008, 1 December 2018, https://web.archive.org/web/20080613055524/http ://latimesblogs.latimes.com:80/technology/2008/06/finally-steve-j.html

44. EWC, "Finally, Steve Jobs Unveils iPhone 2.0 and iPhone 3G," *Los Angeles Times*, 9 June 2008, 1 December 2018, https://web.archive.org/web/20080613055524 /http://latimesblogs.latimes.com:80/technology/2008/06/finally-steve-j.html; Michelle Quinn, "The iPhone Fingernail Problem," *Los Angeles Times*, 12 June 2008, 14 May 2018, http://latimesblogs.latimes.com/technology/2008/06/the-fingernail.html

45. Your Uncle Cordelia, "CAPS LOCK," Urban Dictionary, 16 September 2006, 28 July 2021, https://www.urbandictionary.com/define.php?term=CAPSLOCK

46. Ahmed, *Living A Feminist Life*.

47. Johnson, "Teaching Deconstructively."

48. Michelle Quinn, "Latest Technology Is at Your Fingertips," *Los Angeles Times*, 4 February 2007, 14 May 2018, http://www.latimes.com/la-fi-iphone4-2008feb04 -story.html

49. Michelle Quinn, "The iPhone Fingernail Problem," *Los Angeles Times*, 12 June 2008, 14 May 2018, http://latimesblogs.latimes.com/technology/2008/06/the-finger nail.html

50. Sigmund Freud, "Femininity," in *New Introductory Lectures on Psycho-analysis*, trans. James Strachey (New York: W. W. Norton, 1965), 141.

51. Mary Ann Doane, *Femmes Fatales: Feminism, Film Theory, Psychoanalysis* (New York: Routledge, 1991), 19.

52. Michele White, *The Body and the Screen: Theories of Internet Spectatorship* (Cambridge, MA: MIT Press, 2006).

53. Erica Watson-Currie, as cited in Michelle Quinn, "The iPhone Fingernail Problem," *Los Angeles Times*, 12 June 2008, 11 June 2021, http://latimesblogs.latimes.com/technology/2008/06/the-fingernail.html

54. Erica, "Finally, Steve Jobs Unveils iPhone 2.0 and iPhone 3G," *Los Angeles Times*, 9 June 2008, 1 December 2018, https://web.archive.org/web/20080613055524/http://latimesblogs.latimes.com:80/technology/2008/06/finally-steve-j.html

55. Sara Ahmed, "Cutting Yourself Off," feministkilljoys, 3 November 2017, 1 July 2019, https://feministkilljoys.com/2017/11/03/cutting-yourself-off/

56. Sobieraj and Berry, "From Incivility to Outrage."

57. Michelle Quinn, "The iPhone Fingernail Issue Redux: Adapt or Resist?" *Los Angeles Times*, 25 June 2008, 14 May 2018, http://latimesblogs.latimes.com/technology/2008/06/the-iphone-fing.html

58. Sherifftruman, "Women's Nails vs. iPhone," MacRumors, 5 April 2011, 20 May 2018, https://forums.macrumors.com/threads/womens-nails-vs-iphone.1131322/

59. Oudshoorn and Pinch, *How Users Matter*.

60. Michelle Quinn, "The iPhone Fingernail Issue Redux: Adapt or Resist?" *Los Angeles Times*, 25 June 2008, 14 May 2018, http://latimesblogs.latimes.com/technology/2008/06/the-iphone-fing.html

61. Quinn, "iPhone Fingernail Problem."

62. Quinn, "iPhone Fingernail Issue Redux: Adapt or Resist?"

63. BB user, "The iPhone Fingernail Problem," *Los Angeles Times*, 12 June 2008, 1 December 2018, https://web.archive.org/web/20080613093551/http://latimesblogs.latimes.com/technology/2008/06/the-fingernail.html

64. Dan M., "The iPhone Fingernail Problem," *Los Angeles Times*, 12 June 2008, 1 December 2018, https://web.archive.org/web/20080613093551/http://latimesblogs.latimes.com/technology/2008/06/the-fingernail.html

65. Xtopher, "The iPhone Fingernail Problem," *Los Angeles Times*, 12 June 2008, 1 December 2018, https://web.archive.org/web/20080613093551/http://latimesblogs.latimes.com/technology/2008/06/the-fingernail.html

66. Johnson, "Teaching Deconstructively."

67. Gerard Goggin and Christopher Newell, *Digital Disability* (Lanham, MD: Rowman and Littlefield, 2003).

68. Gerard Goggin, "Disability and Haptic Mobile Media," *New Media and Society* 19, no. 10 (2017): 1571.

69. Carolyn Marvin, *When Old Technologies Were New: Thinking About Electric Communication in the Late Nineteenth Century* (New York: Oxford University Press, 1988).

70. George Kaplan, "The iPhone Fingernail Problem," *Los Angeles Times*, 12 June 2008, 1 December 2018, https://web.archive.org/web/20080613093551/http://latimesblogs .latimes.com/technology/2008/06/the-fingernail.html

71. iPhone user, "The iPhone Fingernail Problem," *Los Angeles Times*, 12 June 2008, 3 March 2022, https://web.archive.org/web/20080613093551/http://latimesblogs .latimes.com/technology/2008/06/the-fingernail.html

72. Shelley Budgeon, "Individualized Femininity and Feminist Politics of Choice," *European Journal of Women's Studies*, 22, no. 3 (2015): 303–18.

73. Linda R. Hirshman, *Get to Work: A Manifesto for Women of the World* (New York: Viking, 2006).

74. Leslie Regan Shade, "Feminizing the Mobile: Gender Scripting of Mobiles in North America," *Continuum: Journal of Media and Cultural Studies* 21, no. 2 (June 2007): 179–89.

75. Microsoft Surface, "Introducing the New Surface Pro 6 – Ultra-light and Versatile," Microsoft, 2018, 5 December 2018, https://www.microsoft.com/en-us/p /surface-pro-6/8ZCNC665SLQ5?ocid=announce_ema_omc_sur_holiday_odo_e1 _mod2_pro&activetab=pivot%3aoverviewtab

76. Dr. GLK, "The iPhone Fingernail Problem," *Los Angeles Times*, 12 June 2008, 1 December 2018, https://web.archive.org/web/20080613093551/http://latimesblogs .latimes.com/technology/2008/06/the-fingernail.html

77. emcee, "The iPhone Fingernail Problem," *Los Angeles Times*, 12 June 2008, 1 December 2018, https://web.archive.org/web/20080613093551/http://latimesblogs .latimes.com/technology/2008/06/the-fingernail.html

78. Oz, "The iPhone Fingernail Problem," *Los Angeles Times*, 12 June 2008, 1 December 2018, https://web.archive.org/web/20080613093551/http://latimesblogs.latimes .com/technology/2008/06/the-fingernail.html

79. Philippa Levine and Alison Bashford, "Introduction: Eugenics and the Modern World," in *The Oxford Handbook of the History of Eugenics*, ed. Alison Bashford and Philippa Levine (New York: Oxford University Press, 2010), 3.

80. RRaya, "The iPhone Fingernail Problem," *Los Angeles Times*, 12 June 2008, 1 December 2018, https://web.archive.org/web/20080613093551/http://latimesblogs .latimes.com/technology/2008/06/the-fingernail.html; Michele White, "How 'your hands look' and 'what they can do': #ManicureMonday, Twitter, and Useful Media," *Feminist Media Histories* 1, no. 2 (Spring 2015): 4–36.

81. Ellen Brandt, "We've Sent You Black Roses and Are Coming to Slaughter Your Pet Hamster: My Life With a Twitter Stalker," EllenInteractive, 14 August 2009, 1 December 2018, https://elleninteractive.wordpress.com/2009/08/03/weve-sent-you-black-roses -and-are-coming-to-slaughter-your-pet-hamster-my-life-with-a-twitter-stalker/

82. Erica Watson-Currie, PhD, "We've Sent You Black Roses and Are Coming to Slaughter Your Pet Hamster: My Life With a Twitter Stalker," EllenInteractive, 14 August 2009, 1 December 2018, https://elleninteractive.wordpress.com/2009/08/03/weve-sent-you -black-roses-and-are-coming-to-slaughter-your-pet-hamster-my-life-with-a-twitter-stalker/

83. s.e. smith, "On Blogging, Threats, and Silence," Tiger Beatdown, 11 October 2011, 11 February 2015, http://tigerbeatdown.com/2011/10/11/on-blogging-threats -and-silence/.

84. Aja Romano, "What We Still Haven't Learned from Gamergate," Vox, 20 January 2020, 16 October 2020, https://www.vox.com/culture/2020/1/20/20808875 /gamergate-lessons-cultural-impact-changes-harassment-laws

85. Lindsay Dogson, "Twitch Streamers Are Sharing Their Stories of Violent Stalkers to Spread Awareness of How to Seek Help," Insider, 18 September 2020, 16 October 2020, https://www.insider.com/twitch-streamers-are-being-stalked-and-harassed-online-2020-9

86. EJ Dickson, "The Short Life and Viral Death of Bianca Devins," *Rolling Stone*, 17 December 2019, 29 November 2020, https://www.rollingstone.com/culture/culture -features/bianca-devins-viral-death-murder-926823/

87. Paul, "Some Women Don't Like iPhones," Ace of Spades HQ, 24 June 2008, 15 May 2018, http://ace.new.mu.nu/some_women_dont_like_iphones. The bold emphasis appears in Paul's text.

88. Robert Franklin, "Amanda Marcotte's still holding out her cup; still no ice cream," A Voice for Men, 15 October 2013, 21 July 2017, https://www.avoiceformen .com/feminism/amanda-marcottes-still-holding-out-her-cup-still-no-ice-cream/

89. Lindell and Sophie, "There Are No Girls on the Internet," Know Your Meme, 7 December 2020, https://knowyourmeme.com/memes/there-are-no-girls-on-the -internet; Adrienne Massanari, "#Gamergate and the Fappening: How Reddit's Algorithm, Governance, and Culture Support Toxic Technocultures," *New Media and Society* 19, no. 3 (2017): 329–46.

90. Nick Farrell, "Apple Fails to Tap Female Market," The Inquirer, 26 June 2008, 14 May 2018, https://www.theinquirer.net/inquirer/news/1048168/apple-hates-women

91. Ken Levine, "The new iPhone hates U," By Ken Levine, 13 July 2008, 14 May 2018, http://kenlevine.blogspot.com/2008/07/new-iphone-hates-u.html

92. Michele White, *Producing Masculinity: The Internet, Gender, and Sexuality* (New York: Routledge, 2019).

93. Amanda Hess, "Why Women Aren't Welcome on the Internet," *Pacific Standard*, 6 January 2014, 7 December 2020, http://www.psmag.com/health-and-behavior /women-arent-welcome-internet-72170

94. alexthechick, "Some Women Don't Like iPhones," Ace of Spades HQ, 24 June 2008, 15 May 2018, http://ace.new.mu.nu/some_women_dont_like_iphones

95. Shoba Sharad Rajgopal, "'The Daughter of Fu Manchu': The Pedagogy of Deconstructing the Representation of Asian Women in Film and Fiction," *Meridians* 10, no. 2 (2010): 141–62.

96. FITCamaro, "Women With Long Fingernails, Fat Fingers Complain About iPhone," DailyTech, 25 June 2008, 22 January 2019, https://web.archive.org/web /20151029135106/http://www.dailytech.com/Women+With+Long+Fingernails+Fat +Fingers+Complain+About+iPhone/article12186c.htm#sthash.LXSqYZL1.dpuf

97. kileil, "Women With Long Fingernails, Fat Fingers Complain About iPhone," DailyTech, 25 June 2008, 22 January 2019, https://web.archive.org/web/20151029135106 /http://www.dailytech.com/Women+With+Long+Fingernails+Fat+Fingers+Complain +About+iPhone/article12186c.htm#sthash.LXSqYZL1.dpuf

98. Tim W., "The new iPhone hates U," By Ken Levine, 14 July 2008, 14 May 2018, http://kenlevine.blogspot.com/2008/07/new-iphone-hates-u.html

99. brian, "Some Women Don't Like iPhones," Ace of Spades HQ, 24 June 2008, 15 May 2018, http://ace.new.mu.nu/some_women_dont_like_iphones

100. Sebastian, "The new iPhone hates U," By Ken Levine, 14 July 2008, 14 May 2018, http://kenlevine.blogspot.com/2008/07/new-iphone-hates-u.html

101. sxr7171, "Women With Long Fingernails, Fat Fingers Complain About iPhone," DailyTech, 26 June 2008, 12 November 2018, https://web.archive.org /web/20151029135106/http://www.dailytech.com/Women+With+Long+Fingern ails+Fat+Fingers+Complain+About+iPhone/article12186c.htm#sthash.LXSqYZL1 .dpuf

102. Angela Sitilides, "I Tried the Long Nail Trend. Here Are My Thoughts." Alexandria Stylebook, 30 April 2019, 15 December 2020, https://alexandriastylebook.com /long-nail-trend-my-thoughts/

103. Carly Cardellino, "15 Things You Super-Annoyingly Can't Do With Long Nails," *Cosmopolitan*, 2 February 2016, 15 December 2020, https://www.cosmo politan.com/style-beauty/beauty/a53074/15-super-annoying-things-you-cant-do -with-long-nails/

104. Noelle Buscher, "iPhone and fingernails," Apple Community, 20 December 2007, 16 May 2018, https://discussions.apple.com/thread/1300856

105. Ken Levine, "The new iPhone hates U," By Ken Levine, 13 July 2008, 14 May 2018, http://kenlevine.blogspot.com/2008/07/new-iphone-hates-u.html

106. Lindawcca, "Screen Touch Control," Apple Community, 23 May 2008, 16 May 2018, https://discussions.apple.com/thread/1532549

107. SolidSmack, "A Touch of Stylus: Touchscreen-Friendly Fingernails Have Arrived!" 17 January 2014, 16 May 2018, https://www.solidsmack.com/culture/touch-stylus -touchscreen-friendly-fingernails-arrived/

108. Amanda Kooser, "Nano Nails Turns Long Fingernails into Touch-screen Styli," CNET, 14 January 2013, 14 May 2018, https://www.cnet.com/news/nano-nails-turns -long-fingernails-into-touch-screen-stylii/

109. Sri Vellanki, as cited in Dean Takahashi, "Elektra Nails let you use your decorated fingernails to tap on a smartphone," Venture Beat, 11 January 2014, 14 May 2018, https://venturebeat.com/2014/01/11/elektra-nails-let-you-use-your-decorated -fingernails-to-tap-on-a-smartphone/

110. teepingpom, "Touchscreen Friendly Fingernails Turn Every Digit into a Stylus," Gizmodo, 15 January 2013, 14 May 2018, https://gizmodo.com/5976066 /touchscreen-friendly-fingernails-turn-every-digit-into-a-stylus

111. Chop-Sue-Me, "Touchscreen Friendly Fingernails Turn Every Digit into a Stylus," Gizmodo, 16 January 2013, 14 May 2018, https://gizmodo.com/5976066 /touchscreen-friendly-fingernails-turn-every-digit-into-a-stylus

112. Sobieraj and Berry, "From Incivility to Outrage."

113. Buster Hein, "Your Long Fingernails Can Now Be Transformed Into Touchscreen Styluses," Cult of Mac, 15 January 2013, 14 May 2018, https://www.cultofmac.com /210391/your-long-fingernails-can-now-be-transformed-into-touchscreen-styluses/

114. Dan Moren, "Long Fingernails and iPhone Typing Don't Mix," *Macworld*, 24 June 2008, 15 May 2018, https://www.macworld.com/article/1134141/iphone_fingernails .html

115. Michele White, *Producing Women: The Internet, Traditional Femininity, Queerness, and Creativity* (New York: Routledge, 2015).

116. Carol Clover, *Men, Women, and Chain Saws: Gender in the Modern Horror Film* (Princeton, NJ: Princeton University Press, 1992).

117. White, "How 'your hands look.'"

118. SolidSmack, "A Touch of Stylus: Touchscreen-Friendly Fingernails Have Arrived!" 17 January 2014, 13 September 2021, https://www.solidsmack.com/culture /touch-stylus-touchscreen-friendly-fingernails-arrived/

119. Emily Blake, "The new iPhone hates U," By Ken Levine, 14 July 2008, 14 May 2018, http://kenlevine.blogspot.com/2008/07/new-iphone-hates-u.html

120. Hope Jahren, Twitter, 18 November 2013, 7 December 2020, https://twitter .com/HopeJahren/status/402471423660150784

121. Allan Sampson, "Iphone and women with long nails," Apple Community, 5 July 2007, 17 May 2018, https://discussions.apple.com/thread/1029547

122. MaximusMushu, "Women's Nails vs. iPhone," MacRumors, 4 April 2011, 1 March 2022, https://forums.macrumors.com/threads/womens-nails-vs-iphone.1131322/;

eastercat, "Women's Nails vs. iPhone," MacRumors, 4 April 2011, 1 March 2022, https://forums.macrumors.com/threads/womens-nails-vs-iphone.1131322/

123. adebaybee, "Multi Touch Display sense long fingernails?" Apple Community, 12 March 2011, 21 August 2021, https://discussions.apple.com/thread/2772385

124. ljahnke, "can the iphone be used with fake nails? If the LG chocolate can do it, I am sure that apple can," Apple Community, 21 September 2011, 17 May 2018, https://discussions.apple.com/thread/3339579

125. Clover, *Men, Women, and Chain Saws*. Clover argues that monsters and final girls fight with less-technological devices, such as knives and fingernails. She identifies the "final girl" as the woman in the film who has a more masculine name, does not have sex, and fights and lives rather than being brutally murdered.

126. Julian Wright, "can the iphone be used with fake nails? If the LG chocolate can do it, I am sure that apple can," Apple Community, 21 September 2011, 17 May 2018, https://discussions.apple.com/thread/3339579

127. Johnson, "Teaching Deconstructively," 142.

128. Annie Kreighbaum, "The Long Nail Life," Into the Gloss, May 2014, 31 December 2018, https://intothegloss.com/2014/05/how-to-grow-long-nails/

129. Tanker Bob, "Samsung t629," Mobile Tech Review, December 2006, 31 December 2018, http://www.mobiletechreview.com/phones/Samsung-t629.htm

130. Wilson Wong, "Review: N76 - A Motorola RAZR wannabe?" Singapore Bikes, 26 June 2007, 30 December 2018, https://www.singaporebikes.com/forums/showthread.php/145678-Review-N76-A-Motorola-RAZR-wannabe?s=e4ac7fdd8da139d4e6328d520bf9a0d8

131. Donna Haraway, *Simians, Cyborgs, and Women: The Reinvention of Nature* (New York: Routledge, 1991).

132. Rommes, "Creating Places for Women."

133. Ahmed, *Living a Feminist Life*.

134. Nomy Bitman and Nicholas A. John, "Deaf and Hard of Hearing Smartphone Users: Intersectionality and the Penetration of Ableist Communication Norms," *Journal of Computer-Mediated Communication* 24 (2019): 56–72.

Chapter 2

1. iphonefreak450, "Cleaning iPhone screen?" MacRumors, 27 July 2017, 20 May 2018, https://forums.macrumors.com/threads/cleaning-iphone-screen.2058794/

2. AmazingTechGeek, "How to avoid fingerprints/smudges from attracting on an iPhone," MacRumors, 15 February 2017, 17 July 2019, https://forums.macrumors

.com/threads/how-to-avoid-fingerprints-smudges-from-attracting-on-an-iphone.2032912/

3. Applejuiced, "How to avoid fingerprints/smudges from attracting on an iPhone," MacRumors, 15 February 2017, 17 July 2019, https://forums.macrumors.com/threads/how-to-avoid-fingerprints-smudges-from-attracting-on-an-iphone.2032912/

4. Newtons Apple, "How to avoid fingerprints/smudges from attracting on an iPhone," MacRumors, 15 February 2017, 17 July 2019, https://forums.macrumors.com/threads/how-to-avoid-fingerprints-smudges-from-attracting-on-an-iphone.2032912/

5. Amy Erdman Farrell, *Fat Shame: Stigma and the Fat Body in American Culture* (New York: New York University Press, 2011), 2.

6. Michele White, *The Body and the Screen: Theories of Internet Spectatorship* (Cambridge, MA: MIT Press, 2006).

7. Sara Ahmed, "The Skin of the Community: Affect and Boundary Formation," in *Revolt, Affect, Collectivity: The Unstable Boundaries of Kristeva's Polis*, ed. Tina Chanter and Ewa Plonowska Ziarek (Albany: State University of New York Press, 2012), 95–111; Didier Anzieu, *The Skin-ego*, trans. Naomi Segal (London: Karnac Books, 2016); Nicolette Bragg, "'Beside Myself': Touch, Maternity and the Question of Embodiment," *Feminist Theory* 21, no. 2 (2020): 141–55; Naomi Segal, "'A Petty Form of Suffering': A Brief Cultural Study of Itching," *Body and Society* 24, nos. 1–2 (2018): 88–102.

8. Steven Connor, *The Book of Skin* (London: Reaktion Books, 2004).

9. Marc Lafrance, "Skin Studies: Past, Present and Future," *Body and Society* 24, nos. 1–2 (2018): 3–32.

10. Lafrance, "Skin Studies," 4.

11. Lafrance, "Skin Studies," 3–4.

12. Lafrance, "Skin Studies," 6.

13. Julia Kristeva, *The Powers of Horror: An Essay on Abjection* (New York: Columbia University Press, 1982).

14. Anzieu, *Skin-ego*, 43.

15. Anzieu, *Skin-ego*, 67.

16. Segal, "Petty Form," 90.

17. Anzieu, *Skin-ego*, 10.

18. Bragg, "Beside Myself," 145.

19. Bragg, "Beside Myself," 147.

20. Ahmed, "Skin of the Community," 101.

21. Ahmed, "Skin of the Community," 101. The italic emphasis appears in Ahmed's text; Apple, "Cleaning your iPhone," 22 October 2020, https://support.apple.com/en-us/HT207123

22. Ahmed, "Skin of the Community," 101.

23. Apple, "Cleaning your iPhone," 22 October 2020, https://support.apple.com/en-us/HT207123

24. Vivian Sobchack, "What My Fingers Knew: The Cinesthetic Subject, or Vision in the Flesh," *Senses of Cinema* 5 (2000): http://sensesofcinema.com/2000/conference-special-effects-special-affects/fingers/. The italic emphasis appears in Sobchack's text.

25. Laura U. Marks, *The Skin of the Film: Intercultural Cinema, Embodiment, and the Senses* (Durham, NC: Duke University Press, 2000), xi.

26. N. Katherine Hayles, *How We Think: Digital Media and Contemporary Technogenesis* (Chicago: University of Chicago Press, 2012).

27. Marks, *Skin of the Film*, xi.

28. Steve Pile, "Spatialities of Skin: The Chafing of Skin, Ego and Second Skins in T. E. Lawrence's *Seven Pillars of Wisdom*," *Body and Society* 17, no. 4: (2011): 65.

29. Lafrance, "Skin Studies."

30. Kristeva, *Powers of Horror*, 2–3.

31. Kristeva, *Powers of Horror*, 101.

32. Naomi Segal, *Consensuality: Didier Anzieu, Gender and the Sense of Touch* (Amsterdam: Rodopi, 2009), 94.

33. Elizabeth Grosz, *Volatile Bodies: Toward a Corporeal Feminism* (Bloomington: Indiana University Press, 1994), 192.

34. David Glover and Cora Kaplan, *Genders* (London: Routledge, 2000).

35. MacNN Staff, "Apple patents improved oleophobic coating process," MacNN, 12 August 2011, 22 October 2016, http://www.macnn.com/articles/11/08/12/could.be.used.in.future.ios.and.mac.products/

36. ajinkya@tmrresearch.com, "Demand for Oleophobic Coatings Market is higher in North America and Europe due to advancements in technologies, developed industries | 2025," Market Reporter, 20 May 2019, 16 June 2019, https://bestmarketherald.com/demand-for-oleophobic-coatings-market-is-higher-in-north-america-and-europe-due-to-advancements-in-technologies-developed-industries-2025/

37. PaulK, "Oleophobic coating – what it is, how to clean your phone, what to do if the coating wears off," PhoneArena, 13 February 2015, 8 June 2019, https://www

.phonearena.com/news/Oleophobic-coating--what-it-is-how-to-clean-your-phone
-what-to-do-if-the-coating-wears-off_id65974

38. Bill Nye, as cited in Wilson Rothman, "Giz Bill Nye Explains: The iPhone 3GS's Oleophobic Screen," Gizmodo, 24 June 2009, 16 June 2019, https://gizmodo.com /giz-bill-nye-explains-the-iphone-3gss-oleophobic-scree-5302097

39. Dictionary of Sexual Terms and Expressions, "waxing the car," 13 July 2019, http://www.sex-lexis.com/-dictionary/waxing+the+car

40. Laura Mulvey, *Visual and Other Pleasures* (Bloomington: Indiana University Press, 1989).

41. Bryan Chaffin, "Bill Nye Explains Apple's Oleophobic iPhone 3GS Screen," iPodObserver, 25 June 2009, 22 October 2016, http://www.ipodobserver.com/ipo /article/bill_nye_explains_apples_oleophobic_iphone_3gs_screen/

42. ch3burashka, "Bill Nye Explains: The iPhone 3GS's Oleophobic Screen," Gizmodo, 24 June 2009, 22 October 2016, http://gizmodo.com/5302097/giz-bill-nye -explains-the-iphone-3gss-oleophobic-screen/

43. wiggin, "Bill Nye Explains: The iPhone 3GS's Oleophobic Screen," Gizmodo, 24 June 2009, 22 October 2016, http://gizmodo.com/5302097/giz-bill-nye-explains-the -iphone-3gss-oleophobic-screen/

44. Manly Housekeeper, "The Proper Way to Clean and Disinfect Your Smartphone," The Manly Housekeeper, 23 October 2012, 22 October 2016, http://www .themanlyhousekeeper.com/2012/10/23/the-proper-way-to-clean-and-disinfect-your -smartphone/

45. Bill Nye, as cited in Wilson Rothman, "Giz Bill Nye Explains: The iPhone 3GS's Oleophobic Screen," Gizmodo, 24 June 2009, 16 June 2019, https://gizmodo.com /giz-bill-nye-explains-the-iphone-3gss-oleophobic-scree-5302097

46. Chris Chavez, "Bring Back the Slick 'New Phone' Feeling to Your Display Using this Amazing Wax," Phandroid, 6 December 2016, 16 June 2019, https://phandroid .com/2016/12/06/android-phone-screen-wax-oleophobic/

47. Rachel Plotnick, "Force, Flatness and Touch Without Feeling: Thinking Historically About Haptics and Buttons," *New Media and Society* 19, no. 10 (2017): 1633.

48. BigDaddy0790, "Think Hard Before Buying an All-glass, Bezel-free Smartphone," The Verge, 29 March 2017, 13 May 2018, https://www.theverge.com/2017/3/29 /15104372/glass-screen-smartphone-design-lg-g6-samsung-galaxy-s8

49. dodger_m, "Oleophobic Screen," Apple Community, 17 August 2009, 1 June 2018, https://discussions.apple.com/thread/2118460

50. aldo82, "Antibacterial phone case," MacRumors, 20 March 2020, https://forums .macrumors.com/threads/antibacterial-phone-case.2227195/?post=28294373#post -28294373

51. Paul Czerwinski, "4 Steps to Properly Clean & Disinfect Mobile Phones, Bear in mind these tips to effectively get the job done." No Jitters, 24 April 2020, 9 May 2020, https://www.nojitter.com/team-collaboration-tools-workspaces/4-steps-properly -clean-disinfect-mobile-phones

52. David Levine, "How to Clean Your Germy Phone," *U.S. News and World Report*, 16 April 2020, 14 July 2021, https://health.usnews.com/conditions/articles/how-to -clean-your-germy-phone

53. Kristeva, *Powers of Horror*.

54. John Harrington and Charles B. Stockdale, "Cleaning Your Devices? Here Are Some Coronavirus Cleaning Tips for Your Phone, Tablet and More," *USA Today*, 5 May 2020, 27 January 2021, https://www.usatoday.com/story/tech/2020/05/05 /coronavirus-cleaning-tips-phone-tablet-devices/111646732/

55. Barbara Johnson, "Teaching Deconstructively," in *Writing and Reading Differently*, ed. George Douglas Atkins, Michael L. Johnson, and Nancy R. Comley (Lawrence: University of Kansas Press, 1986), 140–48.

56. Derek Thompson, "Hygiene Theater Is a Huge Waste of Time," *Atlantic*, 27 July 2020, 21 November 2020, https://www.theatlantic.com/ideas/archive/2020/07 /scourge-hygiene-theater/614599/

57. Juna Xu, "Yikes! You should be cleaning your phone more regularly than you think, *takes out antibacterial wipe*," Body+Soul, 24 April 2020, 14 July 2021, https://www .bodyandsoul.com.au/health/health-advice/yikes-you-should-be-cleaning-your-phone -more-regularly-than-you-think/news-story/a494a9cea45ef07fa5c7b044f2d0b609

58. Kristeva, *Powers of Horror*.

59. Tali Arbel, "Coronavirus: How to Clean the Bundle of Germs that Is Your Phone," *Atlanta Voice*, 15 May 2020, 27 January 2021, https://www.theatlantavoice .com/articles/coronavirus-how-to-clean-the-bundle-of-germs-that-is-your-phone/

60. Josh Ocampo, "Clean Your Phone Right Now," Lifehacker, 1 May 2020, 27 January 2021, https://www.lifehacker.com.au/2020/05/clean-your-phone-right-now/

61. Lisa Gitelman and Geoffrey B. Pingree, "What's New About New Media?" in *New Media, 1740–1915*, ed. Lisa Gitelman and Geoffrey B Pingree (Cambridge, MA: MIT Press, 2003), xviii.

62. Mikhail Bakhtin, *Rabelais and His World*, trans. Hélène Iswolsky (Bloomington: Indiana University Press, 1984).

63. zenpoet, "Bill Nye Explains: The iPhone 3GS's Oleophobic Screen," Gizmodo, 24 June 2009, 22 October 2016, http://gizmodo.com/5302097/giz-bill-nye-explains -the-iphone-3gss-oleophobic-screen/

64. Lafrance, "Skin Studies," 6.

65. See Molly Walker, "Resistant Bacteria Abundant on Nursing Students' Cell Phones," *Medpage Today*, 22 June 2019, https://www.medpagetoday.com/meetingcoverage /asmmicrobe/80657

66. Microsoft at Home and Alyson Munroe, "How to clean your computer," 14 September 2011, http://www.microsoft.com/athome/setup/cleancomputer.aspx

67. Jonathon Millman, as cited in Microsoft at Home and Alyson Munroe, "How to clean your computer," 14 September 2011, http://www.microsoft.com/athome/setup /cleancomputer.aspx

68. Patrick Bass, "Inside a Hard Disk: 'Like Flying a Boeing 747 Six Inches Above the Ground,'" *Antic* 5, no. 6 (October 1986), 18 June 2019, https://www.atarimagazines .com/v5n6/InsideHardDisk.html

69. James Stephen Rutledge, Cory Allen Chapman, Kenneth Scott Seethaler, and William Stephen Duncan, "Systems, Apparatus and Method for Reducing Dust on Components in a Computer System," U.S. Patent 7,113,402, issued 26 September 2006.

70. Keith Evans, "How Does a Computer Monitor Get Dirty?" eHow, 14 September 2011, http://www.ehow.com/how-does_4567285_computer-monitor-dirty .html

71. Bragg, "Beside Myself."

72. Lafrance, "Skin Studies."

73. Anzieu, *Skin-ego*.

74. Fit Moms Fit Kids Club, "Dirty Computer, Phone, or TV Screen? Check out this giveaway!" 7 February 2011, 14 September 2019, http://www.fitmomsfitkidsclub .com/2011/02/dirty-computer-check-out-this-giveaway/

75. Barbara Ehrenreich and Deirdre English, *For Her Own Good: Two Centuries of the Experts' Advice to Women*, 2nd ed. (New York: Random House, 2005), 197. The italic emphasis appears in Ehrenreich and English's text.

76. Michael Bittman, James Mahmud Rice, and Judy Wajcman, "Appliances and Their Impact: The Ownership of Domestic Technology and Time Spent on Household Work," *British Journal of Sociology* 55, no. 3 (2004): 401–23.

77. Feminist historian Ruth Schwartz Cowan describes the increasing criterion for cleanliness expected of modern middle-class women. Ruth Schwartz Cowan, *More Work for Mother: The Ironies of Household Technology from the Open Hearth to the Microwave* (New York: Basic Books, 1983).

78. Jacques Lacan, *The Four Fundamental Concepts of Psycho-analysis*, trans. A. Sheridan (New York: W. W. Norton, 1981).

79. Christian Metz, *The Imaginary Signifier: Psychoanalysis and the Cinema*, trans. Celia Britton, Annwyl Williams, Ben Brewster, and Alfred Guzzetti (Bloomington: Indiana University Press, 1982).

80. Anzieu, *Skin-ego*; Segal, "Petty Form," 90.

81. texasstar, "freakin' hair on camera lens - iPhone X," MacRumors, 3 November 2017, 25 June 2019, https://forums.macrumors.com/threads/freakin-hair-on-camera-lens-iphone-x.2083977/

82. gbrancante, "freakin' hair on camera lens - iPhone X," MacRumors, 15 January 2018, 25 June 2019, https://forums.macrumors.com/threads/freakin-hair-on-camera-lens-iphone-x.2083977/page-4

83. Mary Douglas, *Implicit Meanings: Selected Essays in Anthropology* (London: Routledge, 1999).

84. predation, "freakin' hair on camera lens - iPhone X," MacRumors, 3 November 2017, 25 June 2019, https://forums.macrumors.com/threads/freakin-hair-on-camera-lens-iphone-x.2083977/

85. Kristeva, *Powers of Horror*.

86. Muriel Dimen, "Sexuality and Suffering, or the Eew! Factor," *Studies in Gender and Sexuality* 6, no.1 (2005): 3.

87. martin2345uk, "freakin' hair on camera lens - iPhone X," MacRumors, 3 November 2017, 25 June 2019, https://forums.macrumors.com/threads/freakin-hair-on-camera-lens-iphone-x.2083977/page-2

88. Since "white envelope" products are replacements for damaged phones, they may also be less desirable than the fully packaged device. texasstar, "freakin' hair on camera lens - iPhone X," MacRumors, 3 November 2017, 25 June 2019, https://forums.macrumors.com/threads/freakin-hair-on-camera-lens-iphone-x.2083977/

89. Elena Frank, "Groomers and Consumers: The Meaning of Male Body Depilation to a Modern Masculinity Body Project," *Men and Masculinities* 17, no. 3 (2014): 278–98.

90. UL2RA, "Dust/Speck Inside The iPhone X Camera Lens," MacRumors, 18 November 2017, 11 March 2018, https://forums.macrumors.com/threads/dust-speck-inside-the-iphone-x-camera-lens.2084349/page-5

91. donster28, "Dust/Speck Inside The iPhone X Camera Lens," MacRumors, 6 December 2017, 19 March 2018, https://forums.macrumors.com/threads/dust-speck-inside-the-iphone-x-camera-lens.2084349/page-8

92. Tarja Laine, "Cinema as Second Skin," *New Review of Film and Television Studies* 4, no. 2 (2006): 95.

93. Connor, *Book of Skin.*

94. Bleifuss, "Dirt stuck on the side of screen," Apple Community, 17 July 2008, 24 May 2018, https://discussions.apple.com/thread/1615371

95. rmoliv, "iPhone logo scratched," MacRumors, 20 January 2018, 15 July 2021, https://forums.macrumors.com/threads/iphone-logo-scratched.2102051/#post-25728213

96. Starship67, "iPhone logo scratched," MacRumors, 20 January 2018, 15 July 2021, https://forums.macrumors.com/threads/iphone-logo-scratched.2102051/#post-25728213

97. MacDawg, "iPhone logo scratched," MacRumors, 20 January 2018, 22 May 2018, https://forums.macrumors.com/threads/iphone-logo-scratched.2102051/#post-25728213

98. Sarah Sobieraj and Jeffrey M. Berry, "From Incivility to Outrage: Political Discourse in Blogs, Talk Radio, and Cable News," *Political Communication* 28, no. 1 (2011): 19–41.

99. Sara Ahmed, *Living a Feminist Life* (Durham, NC: Duke University Press, 2017), 21.

100. Remington Steel, "How to avoid fingerprints/smudges from attracting on an iPhone," MacRumors, 15 February 2017, 20 May 2018, https://forums.macrumors.com/threads/how-to-avoid-fingerprints-smudges-from-attracting-on-an-iphone.2032912/

101. MacKid1983, "Bad Smudging - Fingerprints screen," MacRumors, 7 December 2016, 20 May 2018, https://forums.macrumors.com/threads/bad-smudging-fingerprints-screen.2020204/

102. Newtons Apple, "Bad Smudging - Fingerprints screen," MacRumors, 7 December 2016, 20 May 2018, https://forums.macrumors.com/threads/bad-smudging-fingerprints-screen.2020204/

103. White, *Body and the Screen.*

104. johnny_240sx, "Could fingernail cause scratch on rMBP display (or any display)," Apple Community, 29 April 2013, 16 May 2018, https://discussions.apple.com/thread/5004478

105. CCato77, "MacBook Pro 2016 Screen," Apple Community, 3 January 2017, 16 May 2018, https://discussions.apple.com/thread/7813442

106. Lika_tm, "scratched touchpad," Apple Community, 10 June 2017, 20 May 2018, https://discussions.apple.com/thread/7979774

107. Michele White, *Producing Women: The Internet, Traditional Femininity, Queerness, and Creativity* (New York: Routledge, 2015); Michele White, "Women's Nail Polish Blogging and Femininity: 'The Girliest You Will Ever See Me,'" in *Cupcakes, Pinterest,*

and Ladyporn: Feminized Popular Culture in the Early Twenty-First Century, ed. Elana Levine (Urbana-Champaign: University of Illinois Press, 2015), 137–56.

108. jazzdude9792, "Scratch in the Glass," Apple Community, 9 July 2008, 20 May 2018, https://discussions.apple.com/thread/1591128

109. alyabiev, "iPhone X scratched in 4 days," MacRumors, 7 November 2017, 31 March 2018, https://forums.macrumors.com/threads/iphone-x-scratched-in-4-days .2086364/

110. Zach the Apple User, "Scratch on screen," Apple Community, 16 December 2012, 20 May 2018, https://discussions.apple.com/thread/4607260

111. Dimen, "Sexuality and Suffering."

112. kaans, "iPhone X get scratched extremely fast, even with a case," MacRumors, 4 January 2018, 30 March 2018, https://forums.macrumors.com/threads/iphone-x -get-scratched-extremely-fast-even-with-a-case.2086602/page-7

113. Sobchack, "What My Fingers Knew."

114. WISG.1, "When i push on my iPhone screen (5C) it moves in and out," Apple Community, 14 March 2015, 20 May 2018, https://discussions.apple.com/thread /6875445

115. Anne Cranny-Francis, "Semefulness: A Social Semiotics of Touch," *Social Semiotics* 21, no. 4 (September 2011): 473.

116. Nanna Verhoeff, "Theoretical Consoles: Concepts for Gadget Analysis," *Journal of Visual Culture* 8, no. 3 (2009): 279.

117. Roland Barthes, *Camera Lucida: Reflections on Photography*, trans. Richard Howard (New York: Farrar, Strauss, and Giroux, 1981).

118. White, *Body and the Screen.*

119. nasa25, "iPhone X scratched in 4 days," MacRumors, 9 November 2017, 31 March 2018, https://forums.macrumors.com/threads/iphone-x-scratched-in-4-days .2086364/page-4

120. kayzee, "Sony's Xperia Z5 has a frosted glass back, and I've already broken it," The Verge, 8 October 2015, 13 May 2018, https://www.theverge.com/2015/10/8 /9480135/sony-xperia-z5-broken-glass-back

121. ThatiPhoneKid, "Girlfriend Vacuumed my iPhone X," MacRumors, 13 December 2017, 20 June 2019, https://forums.macrumors.com/threads/girlfriend -vacuumed-my-iphone-x.2094759/

122. keysofanxiety, "Girlfriend Vacuumed my iPhone X," MacRumors, 13 December 2017, 20 June 2019, https://forums.macrumors.com/threads/girlfriend-vacuumed -my-iphone-x.2094759/

123. HappyDude20, "Is a Broken iPhone Screen Unprofessional/Embarrassing? (Does it Depend on Factors?)," MacRumors, 14 March 2013, 10 March 2018, https://forums.macrumors.com/threads/is-a-broken-iphone-screen-unprofessional -embarrassing-does-it-depend-on-factors.1556574/

124. scaredpoet, "Is a Broken iPhone Screen Unprofessional/Embarrassing? (Does it Depend on Factors?)," MacRumors, 14 March 2013, 10 March 2018, https://forums .macrumors.com/threads/is-a-broken-iphone-screen-unprofessional-embarrassing -does-it-depend-on-factors.1556574/

125. JohnLT13, "Is a Broken iPhone Screen Unprofessional/Embarrassing? (Does it Depend on Factors?)," MacRumors, 16 March 2013, 10 March 2018, https://forums .macrumors.com/threads/is-a-broken-iphone-screen-unprofessional-embarrassing -does-it-depend-on-factors.1556574/page-2

126. babycake, "Is a Broken iPhone Screen Unprofessional/Embarrassing? (Does it Depend on Factors?)," MacRumors, 15 March 2013, 10 March 2018, https://forums .macrumors.com/threads/is-a-broken-iphone-screen-unprofessional-embarrassing -does-it-depend-on-factors.1556574/page-2

127. Shaun Richardson and Amy Donley, "'That's So Ghetto!' A Study of the Racial and Socioeconomic Implications of the Adjective Ghetto," *Theory in Action* 11, no. 4 (October 2018): 22–43.

128. Kenzo K. Sung, "'Hella Ghetto!' (Dis)locating Race and Class Consciousness in Youth Discourses of Ghetto Spaces, Subjects and Schools," *Race Ethnicity and Education*, 18, no. 3 (2015): 364.

129. iphonefreak450, "Cleaning iPhone screen?" MacRumors, 27 July 2017, 20 May 2018, https://forums.macrumors.com/threads/cleaning-iphone-screen.205 8794/

130. BugeyeSTI, "iPhone logo scratched," MacRumors, 20 January 2018, 22 May 2018, https://forums.macrumors.com/threads/iphone-logo-scratched.2102051/#post -25728213

131. sean000, "Apple Leather Cases - Patina Proud Photos," MacRumors, 4 January 2018, 25 May 2018, https://forums.macrumors.com/threads/apple-leather-cases -patina-proud-photos.2091430/page-5

132. MrMister111, "Which Apple leather is best for not showing dirt/patina?" MacRumors, 4 November 2017, 22 May 2018, https://forums.macrumors.com/threads /which-apple-leather-is-best-for-not-showing-dirt-patina.2084647/

133. Ralfi, "Which Apple leather is best for not showing dirt/patina?" MacRumors, 4 November 2017, 22 May 2018, https://forums.macrumors.com/threads/which -apple-leather-is-best-for-not-showing-dirt-patina.2084647/

134. The italic emphasis appears in Ralfi's text.

135. Ralfi, "Apple Leather Cases - Patina Proud Photos," MacRumors, 29 November 2017, 25 May 2018, https://forums.macrumors.com/threads/apple-leather-cases-patina -proud-photos.2091430/

136. Dimen, "Sexuality and Suffering"; Ralfi, "Apple Leather Cases - Patina Proud Photos," MacRumors, 2 January 2018, 9 May 2022, https://forums.macrumors.com /threads/leather-cases-patina-proud-photos.2091430/page-4

137. chriscrowlee, "Apple Leather Cases - Patina Proud Photos," MacRumors, 2 January 2018, 25 May 2018, https://forums.macrumors.com/threads/apple-leather -cases-patina-proud-photos.2091430/page-4

138. Anzieu, *Skin-ego.*

139. Bragg, "Beside Myself."

140. Ryan1524, "Apple Leather Cases - Patina Proud Photos," MacRumors, 12 January 2018, 22 August 2021, https://forums.macrumors.com/threads/leather-cases -patina-proud-photos.2091430/page-6

141. Molly McHugh, "The Secret Shame of the Cracked iPhone," The Ringer, 29 March 2017, 10 Mary 2018, https://www.theringer.com/2017/3/29/16036460/cracked -iphone-screens-shame-apple-smartphones-9361069088d7

142. Linda Nochlin, *Women, Art, and Power and Other Essays* (New York: Harper and Row, 1984); Griselda Pollock, *Vision and Difference: Femininity, Feminism and the Histories of Art* (London: Routledge, 1988).

143. Lynda Nead, "Seductive Canvases: Visual Mythologies of the Artist and Artistic Creativity," *Oxford Art Journal* 18, no. 2 (1995): 59–69.

144. NBAasDOGG, "Naked iPhone scratches camera bump?" MacRumors, 17 December 2017, 31 March 2018, https://forums.macrumors.com/threads/naked -iphone-scratches-camera-bump.2089463/

145. Jessalynn Keller, "Beware the Dancing Communist: Alexandria Ocasio-Cortez, Snap-lash, and the Politics of Embodied Joy," *Anti-feminisms in Media Culture*, ed. Michele White and Diane Negra (London: Routledge, 2022), 158–75.

146. Sarah Murray, "Postdigital Cultural Studies," *International Journal of Cultural Studies* 23, no. 4 (2020): 445.

147. Ahmed, *Living a Feminist Life.*

148. davidec, "iPhone X in jeans pocket, will screen have micro scratches?" MacRumors, 5 January 2018, 30 March 2018, https://forums.macrumors.com/threads /iphone-x-in-jeans-pocket-will-screen-have-micro-scratches.2085840/page-2

149. iwonder36, "my first screen scratch," Apple Community, 12 July 2007, 17 May 2018, https://discussions.apple.com/thread/1040839

150. wiggin, "Bill Nye Explains: The iPhone 3GS's Oleophobic Screen," Gizmodo, 24 June 2009, 22 October 2016, http://gizmodo.com/5302097/giz-bill-nye-explains -the-iphone-3gss-oleophobic-screen/

151. bigjnyc, "One month later...anyone else's iPhone X still a virgin?" MacRumors, 5 December 2017, 26 March 2018, https://forums.macrumors.com/threads/one -month-later-anyone-elses-iphone-x-still-a-virgin.2093173/

152. Vermifuge, "One month later...anyone else's iPhone X still a virgin?" Mac-Rumors, 9 December 2017, 26 March 2018, https://forums.macrumors.com/threads /one-month-later-anyone-elses-iphone-x-still-a-virgin.2093173/

153. Knowledge Bomb, "One month later...anyone else's iPhone X still a virgin?" MacRumors, 5 December 2017, 26 March 2018, https://forums.macrumors.com /threads/one-month-later-anyone-elses-iphone-x-still-a-virgin.2093173/

154. haqsha23, "iPhone X get scratched extremely fast, even with a case," Mac-Rumors, 9 November 2017, 30 March 2018, https://forums.macrumors.com/threads /iphone-x-get-scratched-extremely-fast-even-with-a-case.2086602/page-5

155. Murray, "Postdigital Cultural Studies."

Chapter 3

1. A preliminary version of my research on heart buttons and narratives was presented at the Society for Cinema and Media Studies Annual Conference in Atlanta, Georgia, in 2016. My thanks to Jason Farman, Hollis Griffin, and Sarah Murray for including me in the "Mediating Mood: Experiencing the Interfaces of Social Media" panel.

2. The Global Language Monitor, "The Heart ♥ Emoji (for love) is Top Word, Pope Francis topped by Ebola as Top Name, 'Hands Up, No Shoot' is Top Phrase," 22 December 2014, 30 November 2020, https://languagemonitor.com/global-english /the-heart-%E2%99%A5-emoji-for-love-is-top-word-pope-francis-topped-by-ebola-as -top-name-hands-up-no-shoot-is-top-phrase/

3. Amy Lee, "LOL, OMG, ♥ Added To The Oxford English Dictionary," *HuffPost*, 25 May 2011, 30 November 2020, https://www.huffpost.com/entry/lol-omg-oxford -english-dictionary_n_840229

4. OED Online, "heart, v." *Oxford English Dictionary*, 3rd ed. (Oxford: Oxford University Press, 2004), March 2016, 30 November 2020, http://www.oed.com/view /Entry/85069?rskey=JB4IhQ&result=1

5. Alexandra Petri, "Stop it, Oxford English Dictionary. OMG? LOL? Heart? No!" *Washington Post*, 24 March 2011, 30 November 2020, https://www.washingtonpost .com/blogs/compost/post/stop-it-oxford-english-dictionary-omg-lol-heart-no/2011 /03/03/ABLH9ARB_blog.html

6. Jonny Evans, "Apple Has a Chance to Build a Social Network We Can Trust," *Computerworld*, 21 May 2019, 29 May 2020, https://www.computerworld.com/article /3396996/apple-has-a-chance-to-build-a-social-network-we-can-trust.html

7. Bill Browning, "Dad Goes Viral for Tweeting About Trans Son's Party to Celebrate Transition: The Love that Shines through It All Has Captured the Heart of Social Media," *LGBTQ Nation*, 24 February 2020, 29 May 2020, https://www.lgbtqnation .com/2020/02/dad-goes-viral-tweeting-trans-sons-party-celebrate-transition/; David Rasbach, "CNN Viewers Went Crazy for this Cat. She's from Bellingham," *Bellingham Herald*, 28 October 2019, 29 May 2020, https://www.bellinghamherald.com /news/local/crime/article236733028.html

8. Northshore Veterinary Hospital, "Jason encourages Cinder! That's Good Work!" Facebook, 19 October 2019, 1 June 2020, https://www.facebook.com/watch/?ref =external&v=942862712839651

9. Apple, "iPad Pro," 13 January 2016, http://www.apple.com/ipad-pro/#product-demo

10. Etsy, "About Etsy," 1 September 2016, https://www.etsy.com/about

11. Michele White, *Producing Women: The Internet, Traditional Femininity, Queerness, and Creativity* (New York: Routledge, 2015).

12. Steve Woolgar, "Configuring the User: The Case of Usability Trials," in *A Sociology of Monsters: Essays on Power, Technology and Domination*, ed. John Law (London: Routledge, 1991), 58–99.

13. Roland Barthes, *Camera Lucida: Reflections on Photography*, trans. Richard Howard (New York: Farrar, Strauss, and Giroux, 1981).

14. Kevin Roberts, *Lovemarks: The Future Beyond Brands* (New York: PowerHouse Books, 2004); Brody J. Ruihley and Joshua R. Pate, "For the Love of Sport: Examining Sport Emotion Through a Lovemarks Lens," *Communication and Sport* 5, no. 2 (2017): 135–59.

15. Barthes, *Camera Lucida*, 26.

16. Elspeth H. Brown and Thy Phu, "Introduction," in *Feeling Photography*, ed. Elspeth H. Brown and Thy Phu (Durham, NC: Duke University Press, 2014), 1–28.

17. Shawn Michelle Smith, "Photography Between Desire and Grief: Roland Barthes and F. Holland Day," in *Feeling Photography*, ed. Elspeth H. Brown and Thy Phu (Durham, NC: Duke University Press, 2014), 30.

18. Barthes, *Camera Lucida*, 27.

19. Barthes, *Camera Lucida*, 26–27.

20. Roland Barthes, *Incidents*, trans. Richard Howard (Berkeley: University of California Press, 1992).

21. Barthes, *Camera Lucida*, 27. The italic emphasis appears in Barthes's text.

22. Barthes, *Camera Lucida*, 28.

23. Woolgar, "Configuring the User," 58. The italic emphasis appears in Woolgar's text.

24. Nigel Thrift, "Closer to the Machine? Intelligent Environments, New Forms of Possession and the Rise of the Supertoy," *Cultural Geographies* 10 (2003): 389–407.

25. Anne Sofie Lægran and James Stewart, "Nerdy, Trendy, or Healthy? Configuring the Internet Café," *New Media and Society* 5, no. 3 (2003): 357–77.

26. Madeleine Akrich, "The De-scription of Technical Objects," in *Shaping Technology/Building Society: Studies in Sociotechnical Change*, ed. Wiebe E. Bijker and John Law (Cambridge, MA: MIT Press, 1994), 205–24; Majken Kirkegaard Rasmussen and Marianne Graves Petersen, "Re-scripting Interactive Artefacts with Feminine Values," in *Proceedings of the 2011 Conference on Designing Pleasurable Products and Interfaces* (Milan: ACM, 2011), 1–8; Nelly Oudshoorn, Els Rommes, and Marcelle Stienstra, "Configuring the User as Everybody: Gender and Design Cultures in Information and Communication Technologies," *Science, Technology, and Human Values* 29, no. 1 (Winter 2004): 30–63.

27. Mary Ann Doane, *Femmes Fatales: Feminism, Film Theory, Psychoanalysis* (New York: Routledge, 1991).

28. Laura Mulvey, *Visual and Other Pleasures* (Bloomington: Indiana University Press, 1989).

29. Armin Dietz, "Heart Symbol & Heart Burial," 9 May 2019, http://www.heartsymbol.com/english/index.html

30. Theodore V. Buttrey, "The Coins and the Cult," *Expedition Magazine* 34, nos. 1–2 (Spring–Summer 1992): 59–68.

31. P. J. Vinken, *The Shape of the Heart* (Amsterdam: Elsevier, 2000).

32. Charles Knutson and MacGregor Games, "A Breef History of Playing Cardes," *Renaissance Magazine*, 2001, 4 March 2016, http://www.renaissancemagazine.com/backissues/game.html

33. Keelin McDonald, "The Shape of My Heart: Where Did the Ubiquitous Valentine's Symbol Come From?" *Slate*, February 2007, 20 March 2016, http://www.slate.com/articles/news_and_politics/recycled/2007/02/the_shape_of_my_heart.html?GT1=9129

34. Herbert Smith, "Badges, Buttons, T-shirts and Bumperstickers: The Semiotics of Some Recursive Systems," *Journal of Popular Culture* 21, no. 4 (1988): 141–49.

35. John F. Sherry and Mary Ann McGrath, "Unpacking the Holiday Presence: A Comparative Ethnography of Two Gift Stores," *Interpretive Consumer Research* (1989):

148–67, http://futureofchildren.acrwebsite.org/search/view-conference-proceedings.aspx?Id=12182

36. Noel Albert and Dwight Merunka, "The Role of Brand Love in Consumer-Brand Relationships," *Journal of Consumer Marketing* 30, no. 3 (2013): 258–66.

37. Barbara A. Carroll and Aaron C. Ahuvia, "Some Antecedents and Outcomes of Brand Love," *Marketing Letters* 17 (2006): 80.

38. Carroll and Ahuvia, "Some Antecedents and Outcomes of Brand Love," 81.

39. Susanna Paasonen, Ken Hillis, and Michael Petit, "Introduction: Networks of Transmission: Intensity, Sensation, Value," in *Networked Affect*, ed. Ken Hillis, Susanna Paasonen, and Michael Petit (Cambridge, MA: MIT Press, 2015), 1.

40. Emojipedia, "Red Heart," 27 March 2019, https://emojipedia.org/heavy-black-heart/

41. Stormfront, "Stormfront - White Nationalist Community," 24 May 2019, https://www.stormfront.org/forum/

42. Stormfront, "Stormfront - Smilies," 24 May 2019, https://www.stormfront.org/forum/misc.php?do=showsmilies

43. Barthes, *Camera Lucida*, 26.

44. For a discussion of the affective aspects of emojis, see Luke Stark and Kate Crawford, "The Conservatism of Emoji: Work, Affect, and Communication," *Social Media + Society* (July–December 2015): 1–11.

45. Lauren Collister, "There's a reason using a period in a text message makes you sound angry," Quartz, 3 January 2018, 31 July 2019, https://qz.com/1169792/theres-a-reason-using-a-period-in-a-text-message-makes-you-sound-angry/

46. Ariadna Matamoros-Fernández, "Inciting Anger Through Facebook Reactions in Belgium: The Use of Emoji and Related Vernacular Expressions in Racist Discourse," *First Monday* 23, no. 9, 3 September 2018, 23 May 2019, https://www.firstmonday.org/ojs/index.php/fm/article/view/9405

47. Scott E. Fahlman, as cited in "Original Bboard Thread in which :-) was proposed," 19 September 1982, 3 February 2021, http://www.cs.cmu.edu/%7Esef/Orig-Smiley.htm

48. Eric Raymond, ed., "Revision History," The Jargon File, 11 December 2020, http://www.catb.org/~esr/jargon/html/revision-history.html; Guy L. Steele, ed., *The Hacker Dictionary* (New York: Harper and Row, 1983).

49. Hacker's Dictionary, "emoticon," 1 March 2022, http://hackersdictionary.com/html/The-Jargon-Lexicon-framed.html

50. Paul Andrews, "Put On A Happy Face, But Not In My E-Mail." *Seattle Times*, 1994, 14 April 2018, http://www.mit.edu/people/cordelia/smileys_edit.html

51. Emojipedia, "Home of Emoji Meanings," 10 April 2019, https://emojipedia.org/

52. Dictionary.com, "What does ♥ Black Heart Emoji mean?" 1 June 2020, https://www.dictionary.com/e/emoji/black-heart-emoji/

53. Dictionary.com, "What does the red heart emoji mean?" 10 April 2019, http://www.dictionary.com/e/emoji/red-heart-emoji/

54. Dictionary.com, "What does the red heart emoji mean?" 2 May 2018, http://www.dictionary.com/e/emoji/red-heart-emoji/

55. Aly Trachtman, "What Your Emoji Heart Color of Choice Says About You Why use green? WHY?" FlockU, 2 May 2018, http://flocku.com/articles/what-your-emoji-heart-color-of-choice-says-about-you

56. Steven L. Arxer, "Hybrid Masculine Power: Reconceptualizing the Relationship Between Homosociality and Hegemonic Masculinity," *Humanity and Society* 35 (November 2011): 390–422.

57. craigums, "<3," Urban Dictionary, 6 March 2005, 10 April 2019, https://www.urbandictionary.com/define.php?term=%3C3&page=3. For a discussion of misogynistic practices on Urban Dictionary, see Debbie Ging, Theodore Lynn, and Pierangelo Rosati, "Neologising Misogyny: Urban Dictionary's Folksonomies of Sexual Abuse," *New Media and Society* 22, no. 5 (2020): 838–56.

58. craigums, "lyke omg," Urban Dictionary, 4 December 2005, 10 April 2019, https://www.urbandictionary.com/define.php?term=lyke+omg&defid=1541263

59. Elizabeth Losh, *Hashtag* (New York: Bloomsbury Publishing, 2019).

60. Shitbrick1502, "Hashtag," Urban Dictionary, 13 June 2013, 24 January 2021, https://www.urbandictionary.com/define.php?term=hashtag; RaggedCoyote, "Hashtag," Urban Dictionary, 20 May 2014, 24 January 2021, https://www.urbandictionary.com/define.php?term=hashtag&page=2

61. Birdypwnsjoo, "<3," Urban Dictionary, 12 May 2006, 14 April 2018, http://www.urbandictionary.com/define.php?term=%3C3&page=4

62. Ulrecht, "<3," Urban Dictionary, 11 May 2010, 30 January 2021, https://www.urbandictionary.com/define.php?term=%3C3&page=8

63. Susan Brownmiller, *Femininity* (New York: Simon and Schuster, 1984); Sandra L. Bartky, "Foucault, Femininity, and the Modernization of Patriarchal Power," in *Feminism and Foucault*, ed. Irene Diamond and Lee Quinby (Boston: Northeastern University Press, 1988), 61–86.

64. Adele Patrick, "Defiantly Dusty: A (Re)Figuring of 'Feminine Excess,'" *Feminist Media Studies* 1, no. 3 (2001): 361.

65. Sara Ahmed, *Living a Feminist Life* (Durham, NC: Duke University Press, 2017).

66. Sarah Banet-Weiser and Kate M. Miltner, "#MasculinitySoFragile: Culture, Structure, and Networked Misogyny," *Feminist Media Studies* 16, no. 1 (2016): 171–74.

67. Andre, "<3," Urban Dictionary, 19 February 2005, 14 April 2018, http://www.urbandictionary.com/define.php?term=%3C3&page=13

68. Michele White, *Buy It Now: Lessons from eBay* (Durham, NC: Duke University Press, 2012); White, *Producing Women*.

69. Christoph Fuchs, Martin Schreier, and Stijn M. J. van Osselaer, "The Handmade Effect: What's Love Got to Do with It?" *Journal of Marketing* 79 (March 2015): 100.

70. Etsy, "Etsy – Email Sign Up," 12 September 2021, http://www.etsy.com/emails/weddings; White, *Producing Women*.

71. White, *Buy It Now*.

72. Etsy, "Etsy – Shop for handmade, vintage, custom, and unique gifts for everyone," 2 June 2020, https://www.etsy.com/?ref=lgo

73. Diana Adis Tahhan, "Touching at Depth: The Potential of Feeling and Connection," *Emotion, Space and Society* 7 (2013): 45–53.

74. White, *Producing Women*.

75. Etsy, "Etsy – Shop for handmade, vintage, custom, and unique gifts for everyone," 2 June 2020, https://www.etsy.com/?ref=lgo

76. Tahhan, "Touching at Depth," 51.

77. Tahhan, "Touching at Depth," 46.

78. Etsy, "About Etsy," 12 September 2021, https://www.etsy.com/about

79. Lisa Whitmer, "How to Make a Shop Video With Your Smartphone," Etsy, 19 January 2016, 18 February 2019, https://www.etsy.com/seller-handbook/article/how-to-make-a-shop-video-with-your/37195025787

80. Louisa, "How to Make a Shop Video With Your Smartphone," Etsy, 19 January 2016, 18 February 2019, https://www.etsy.com/seller-handbook/article/how-to-make-a-shop-video-with-your/37195025787

81. Adriana, "How to Make a Shop Video With Your Smartphone," Etsy, 20 January 2016, 18 February 2019, https://www.etsy.com/seller-handbook/article/how-to-make-a-shop-video-with-your/37195025787

82. Rachel Plotnick, "Force, Flatness and Touch Without Feeling: Thinking Historically About Haptics and Buttons," *New Media and Society* 19, no. 10 (2017): 1633.

83. Heather Burkman, "Favorite Lists: Collect to Your Heart's Content," Etsy, 25 June 2013, 15 January 2014, http://www.etsy.com/blog/news/2013/favorite-lists -collect-to-your-hearts-content/

84. dfalv38, "100 Etsy Hearts and If You Need A Black Heart... ♥," Iris & Lily Insights, 21 August 2008, 31 May 2020, https://irisandlilydesigns.wordpress.com/2008/08/21 /100-etsy-hearts-and-if-you-need-a-black-heart-%e2%99%a5/

85. EtsyGadget, "Etsy Hearts Counter," ETSY GADGET, 16 August 2016, http:// etsygadget.com/en/etsy_hearts_counter/

86. Rose Barbola, "Etsy Hearts Counter," ETSY GADGET, 8 April 2016, 16 August 2016, http://etsygadget.com/en/etsy_hearts_counter/

87. Pamela Quinn from Vintagequinngifts, "Etsy Hearts Counter," ETSY GADGET, 6 April 2016, 16 August 2016, http://etsygadget.com/en/etsy_hearts_counter/

88. Val Hebert, "Handmade Needlefelted Goodness by Val Hebert by ValArtsStudio," Etsy, 18 February 2019, https://www.etsy.com/shop/ValsArtStudio#about

89. Etsy, "Etsy – Shop for handmade, vintage, custom, and unique gifts for everyone," 2 June 2020, https://www.etsy.com/?ref=lgo

90. Val Hebert, "Val Hebert on Etsy," Etsy, 18 February 2019, https://www.etsy.com /people/ValsArtStudio

91. Val Hebert, "Handmade Needlefelted Goodness by Val Hebert by ValArtsStudio," Etsy, 12 March 2022, https://www.etsy.com/shop/ValsArtStudio#about

92. Barbara Johnson, "Teaching Deconstructively," in *Writing and Reading Differently*, ed. George Douglas Atkins, Michael L. Johnson, and Nancy R. Comley (Lawrence: University of Kansas Press, 1986), 140–48.

93. Didier Anzieu, *The Skin-ego*, trans. Naomi Segal (London: Karnac Books, 2016).

94. Steven Connor, *The Book of Skin* (London: Reaktion Books, 2004); Tarja Laine, "Cinema as Second Skin," *New Review of Film and Television Studies* 4, no. 2 (2006): 93–106.

95. Elvira Para, "Elvira Para on Etsy," Etsy, 17 August 2016, https://www.etsy.com /people/elvirapara

96. Val Hebert, "Handmade Needlefelted Goodness by Val Hebert by ValArtsStudio," Etsy, 18 February 2019, https://www.etsy.com/shop/ValsArtStudio#about

97. The Heartmaker, "Uniquely Handcrafted Fine Wooden Jewelry & by Heartistics on Etsy," Etsy, 16 August 2016, https://www.etsy.com/shop/Heartistics?ref=ss_profile

98. Beverly Thomas Jenkins, "Beverly Thomas Jenkins on Etsy," Etsy, 18 February 2019, https://www.etsy.com/people/beverlytjenkins

99. The Heartmaker, "Uniquely Handcrafted Fine Wooden Jewelry & by Heartistics on Etsy," Etsy, 16 August 2016, https://www.etsy.com/shop/Heartistics?ref=ss_profile

100. The Heartmaker, "The heartmaker on Etsy," Etsy, 16 August 2016, https://www.etsy.com/people/Heartistics

101. clburon1234, "Uniquely Handcrafted Fine Wooden Jewelry & by Heartistics on Etsy," Etsy, 6 March 2018, 12 February 2019, https://www.etsy.com/shop/Heartistics?ref=ss_profile

102. Caitlin Dewey, "A Quick, No-nonsense Guide to Using Facebook's New Reactions," *Washington Post*, 24 February 2016, 13 April 2019, https://www.washingtonpost.com/news/the-intersect/wp/2016/02/24/a-quick-no-nonsense-guide-to-using-facebooks-new-reactions/?utm_term=.dd0dd5464d30

103. Lin Qiu, Han Lin, Angela K. Leung, and William Tov, "Putting Their Best Foot Forward: Emotional Disclosure on Facebook," *Cyberpsychology, Behavior, and Social Networking* 15, no. 10 (2012): 569–72.

104. Corina Sas, Alan Dix, Jennefer Hart, and Ronghui Su, "Dramaturgical Capitalization of Positive Emotions: The Answer for Facebook Success?" in *Proceedings of the 23rd British HCI Group Annual Conference on People and Computers: Celebrating People and Technology* (London: British Computer Society, 2009), 120.

105. Geoff Teehan, "Reactions: Not everything in life is Likable," Medium, 24 February 2016, 13 April 2018, https://medium.com/facebook-design/reactions-not-everything-in-life-is-likable-5c403de72a3f

106. Fidji Simo, as cited in Evelyn Lau, "Facebook Launches New 'Hug' Reaction Button in Response to Coronavirus," *The National*, 17 April 2020, 31 May 2020, https://www.thenational.ae/lifestyle/facebook-launches-new-hug-reaction-button-in-response-to-coronavirus-1.1007243

107. Alexandru Voica, Twitter, 17 April 2020, 31 May 2020, https://twitter.com/alexvoica/status/1251090864878821377

108. Robinson Meyer, "Facebook Isn't Just Getting a Dislike Button," *Atlantic*, 8 October 2015, 14 April 2018, https://www.theatlantic.com/technology/archive/2015/10/facebook-isnt-just-getting-a-dislike-button/409678/

109. Mark Zuckerberg, as cited in Meyer, "Facebook Isn't Just Getting a Dislike Button."

110. Albert M. Muñiz and Thomas C. O'Guinn, "Brand Community," *Journal of Consumer Research* 27, no. 4 (March 2001): 412–32.

111. Facebook Guidelines, 13 April 2018, https://en.facebookbrand.com/assets/reactions

112. Aura Bogado, "How White Separatists Disable Native American Facebook Accounts Facebook's 'authentic names' policy makes sense in theory," Colorlines, 11 March 2015, 17 April 2019, https://www.colorlines.com/articles/how-white -separatists-disable-native-american-facebook-accounts; Sam Levin, "As Facebook Blocks the Names of Trans Users and Drag Queens, This Burlesque Performer Is Fighting Back," *Guardian*, 29 June 2017, 17 April 2019, https://www.theguardian .com/world/2017/jun/29/facebook-real-name-trans-drag-queen-dottie-lux

113. Electronic Frontier Foundation, "Open Letter to Facebook About its Real Names Policy," 5 October 2015, 17 April 2019, https://www.eff.org/document/open -letter-facebook-about-its-real-names-policy.

114. Shibani Mahtani, "Facebook Removed 1.5 Million Videos of the Christchurch Attacks Within 24 hours—and There Were Still Many More," *Washington Post*, 17 March 2019, 4 August 2021, https://www.washingtonpost.com/world/facebook -removed-15-million-videos-of-the-christchurch-attacks-within-24-hours--and-there -were-still-many-more/2019/03/17/fe3124b2-4898-11e9-b871-978e5c757325_story .html

115. VICE, "Charlottesville: Race and Terror – VICE News Tonight on HBO," You-Tube, 14 August 2017, 23 July 2019, https://www.youtube.com/watch?v=RIrcB1sAN8I

116. Matamoros-Fernández, "Inciting Anger."

117. Andreas Chatzidakis, Jamie Hakim, Jo Littler, Catherine Rottenberg, and Lynne Segal, "From Carewashing to Radical Care: The Discursive Explosions of Care During Covid-19," *Feminist Media Studies* 20, no. 6 (2020): 891.

118. Sammi Krug, "What the Reactions Launch Means for News Feed," Facebook, 24 February 2016, 11 April 2019, https://newsroom.fb.com/news/2016/02/news-feed -fyi-what-the-reactions-launch-means-for-news-feed/

119. Caitlin Dewey, "Facebook's News Feed and the Tyranny of 'Positive' Content," *Washington Post*, 30 June 2016, 13 April 2019, https://www.washingtonpost .com/news/the-intersect/wp/2016/06/30/facebook-news-feed-and-the-tyranny-of -positive-content/?utm_term=.31b06337b8c9

120. Vindu Goel, "Facebook Tinkers with Users' Emotions in News Feed Experiment, Stirring Outcry," *New York Times*, 29 June 2014, 13 April 2019, https://www .nytimes.com/2014/06/30/technology/facebook-tinkers-with-users-emotions-in -news-feed-experiment-stirring-outcry.html

121. Connie Fredrickson, "Every time it's acceptable to use the new Facebook reactions," The Tab, 29 February 2016, 12 September 2021, https://thetab.com/uk/hull /2016/02/29/how-to-use-the-new-facebook-buttons-10059

122. The Smart 1, "vanilla like," Urban Dictionary, 31 March 2019, 13 April 2019, https://www.urbandictionary.com/define.php?term=vanilla%20like

123. Caitlin Dewey, "A Quick, No-nonsense Guide to Using Facebook's New Reactions," *Washington Post*, 24 February 2016, 13 April 2019, https://www .washingtonpost.com/news/the-intersect/wp/2016/02/24/a-quick-no-nonsense -guide-to-using-facebooks-new-reactions/?utm_term=.dd0dd5464d30

124. Brian Adam, "How to Activate the New Reaction of 'I care' on Facebook?" Euro X live, 25 May 2020, https://intallaght.ie/how-to-activate-the-new-reaction-of-i-care -on-facebook/

125. Natt Garun, "Facebook reactions have now infiltrated comments," The Verge, 3 May 2017, 13 April 2019, https://www.theverge.com/2017/5/3/15536812/facebook -reactions-now-available-comments

126. Plotnick, "Force, Flatness."

127. Frank Bank, "Reactions: Not everything in life is Likable," Medium, 25 February 2016, 13 April 2018, https://medium.com/facebook-design/reactions-not-everything -in-life-is-likable-5c403de72a3f

128. N. Katherine Hayles, *How We Think: Digital Media and Contemporary Technogenesis* (Chicago: University of Chicago Press, 2012).

129. Jylian Russell, "Facebook Reactions: What They Are and How They Impact the Feed," Hootsuite, 13 April 2018, https://blog.hootsuite.com/how-facebook-reactions -impact-the-feed/

130. Karissa Bell, "You might want to rethink what you're 'liking' on Facebook now," Mashable, 27 February 2017, 13 April 2018, https://mashable.com/2017/02 /27/facebook-reactions-news-feed/#5DJjTGG1Kkq1

131. Felicity Wild, "How to interpret Facebook reactions," Digital communications team blog, 16 March 2016, 13 April 2019, https://digitalcommunications.wp.st -andrews.ac.uk/2016/03/16/how-to-interpret-facebook-reactions/

132. mroth, "emojitracker: realtime emoji use on twitter," 3 June 2020, http:// emojitracker.com/

133. Twitter offered the star button starting in 2006. Casey Newton, "Twitter officially kills off favorites and replaces them with likes," The Verge, 3 November 2015, 22 April 2019, https://www.theverge.com/2015/11/3/9661180/twitter-vine-favorite -fav-likes-hearts

134. Twitter, 3 November 2015, 12 March 2016, https://twitter.com/twitter/status /661558661131558915?ref_src=twsrctfw

135. Akarshan Kumar, "Hearts on Twitter," Twitter, 3 November 2015, 12 March 2016, https://blog.twitter.com/2015/hearts-on-twitter

136. Fidji Simo, as cited in Evelyn Lau, "Facebook Launches New 'Hug' Reaction Button in Response to Coronavirus," *The National*, 17 April 2020, 31 May 2020,

https://www.thenational.ae/lifestyle/facebook-launches-new-hug-reaction-button
-in-response-to-coronavirus-1.1007243

137. Charlie Warzel, "Why We Favorite Tweets, According To Science," *BuzzFeed*,
13 May 2014, 10 January 2021, https://www.buzzfeed.com/charliewarzel/why-we
-favorite-tweets-according-to-science?utm_term=.agy2Gd0RQ#.ewKx3KOZm

138. Thomas Ricker, "First Click: Twitter's battle between hearts and stars is a battle
for hearts and minds," The Verge, 29 July 2015, 7 April 2018, https://www.theverge
.com/2015/7/29/9065493/twitter-replacing-stars-with-hearts-favorites

139. PiotrNowinski, Twitter, 3 November 2015, 12 March 2016, https://twitter.com
/PiotrNowinski/status/661668448372711425?replies_view=true&cursor=AeDWx9i4Lgk

140. Mashable, Twitter, 3 November 2015, 7 April 2018, https://twitter.com/mashable
/status/661562775030906880

141. npralltech, Twitter, 3 November 2015, 7 April 2018, https://twitter
.com/npralltech/status/661568379740954624?ref_src=twsrc%5Etfw&ref_url
=https%3A%2F%2Fwww.npr.org%2Fsections%2Fthetwo-way%2F2015%2F11%2F03
%2F454368845%2Fwhy-no-love-for-twitters-hearts&tfw_site=NPR

142. jamesoreilly, Twitter, 3 November 2015, 7 April 2018, https://twitter.com
/jamesoreilly/status/661687693101821952

143. Abhimanyu Ghoshal, "Why I don't 'like' Twitter's new 'heart' button," TNW,
4 November 2015, 7 April 2018, https://thenextweb.com/twitter/2015/11/04/why-i
-dont-like-twitters-new-heart-button/

144. Mongoosebumpkin, "Twitter Is Replacing Favourites with Likes—But Does
Anyone Heart It?" *Guardian*, 5 November 2015, 8 April 2018, https://www.theguardian
.com/discussion/p/4dppv

145. Sarah Sobieraj and Jeffrey M. Berry, "From Incivility to Outrage: Political Discourse in Blogs, Talk Radio, and Cable News," *Political Communication* 28, no. 1 (2011):
19–41.

146. BrownoftheGlobe, Twitter, 3 November 2015, 7 April 2018, https://twitter.com
/BrownoftheGlobe/status/661624039014797312

147. CherokeeLair, Twitter, 3 November 2015, 7 April 2018, https://twitter.com
/CherokeeLair/status/661647973873512448

148. katecrawford, Twitter, 3 November 2015, 23 September 2021, https://twitter
.com/katecrawford/status/661570552801443840

149. Mario Aguilar, "Twitter Is Replacing the Fav Star With a Stupid Fucking Heart,"
Gizmodo, 3 November 2015, 8 May 2019, https://gizmodo.com/twitter-is-replacing
-the-fav-star-with-a-stupid-fucking-1740270018. The italic emphasis appears in
Aguilar's text.

150. nazirology, Twitter, 4 November 2015, 7 April 2018, https://twitter.com/nazirology/status/661787188154142720

151. ackraemer, Twitter, 3 November 2015, 12 March 2016, https://twitter.com/ackraemer/status/661580197804826624

152. emilybell, Twitter, 3 November 2015, 7 April 2018, https://twitter.com/emilybell/status/661586412500860928

153. dkiesow, Twitter, 3 November 2015, 26 May 2019, https://twitter.com/dkiesow/status/661587280256040961

154. Mathew Ingram, "Hearts and Faves: How Much Should Twitter Care About its Core Users?" *Fortune*, 5 November 2015, 7 April 2018, http://fortune.com/2015/11/03/twitter-hearts/

Chapter 4

1. Peace and Saraity ASMR, "ASMR| Tapping on Cellphone Cases| iPhone X Case Collection," YouTube, 2 January 2018, 21 July 2019, https://www.youtube.com/watch?v=hwUknNkXiyU. The graphical aspects of member names, video titles, and posts have been retained as much as possible in quotes and citations in this and other chapters. However, some of these elements appear differently in this print version, including appearing in black and white, or were not translatable.

2. Be You, "ASMR| Tapping on Cellphone Cases| iPhone X Case Collection," YouTube, 3 March 2022, https://www.youtube.com/watch?v=hwUknNkXiyU

3. Emma Leigh Waldron, "'This FEELS SO REAL!' Sense and Sexuality in ASMR Videos," *First Monday* 22, nos. 1–2, January 2017, 28 December 2018, https://firstmonday.org/ojs/index.php/fm/article/view/7282/5804

4. Roland Barthes, *Camera Lucida: Reflections on Photography*, trans. Richard Howard (New York: Farrar, Strauss, and Giroux, 1981).

5. S. S., "ASMR| Tapping on Cellphone Cases| iPhone X Case Collection," YouTube, 21 July 2019, https://www.youtube.com/watch?v=hwUknNkXiyU

6. Sara Ahmed, *Living a Feminist Life* (Durham, NC: Duke University Press, 2017).

7. Michele White, "Television and Internet Differences by Design: Rendering Liveness, Presence, and Lived Space," *Convergence: The International Journal of Research into New Media Technologies* 12, no. 3 (August 2006): 341–55.

8. Jane Feuer, *The Hollywood Musical,* 2nd edition (Bloomington: Indiana University Press, 1993), 36.

9. Michele Hilmes, "The Television Apparatus: Direct Address," *Journal of Film and Video* 37, no. 4 (Fall 1985): 28.

10. John Ellis, *Visible Fictions: Cinema: Television: Video* (London: Routledge, 2002), 59.

11. Ellis, *Visible Fictions*, 139.

12. Matthew Lombard and Jennifer Snyder-Duch, "Interactive Advertising and Presence," *Journal of Interactive Advertising* 1, no. 2 (2001): 56–65.

13. John Langer, "Television's 'Personality System,'" *Media, Culture and Society* 3, no. 4 (1981): 364.

14. Victor J. Viser, "Mode of Address, Emotion and Stylistics: Images of Children in American Magazine Advertising, 1940–1950," *Communication Research* 24, no. 1 (February 1997): 83–101.

15. Paul Frosh, "The Face of Television," *Annals of the American Academy* 625 (September 2009): 87–102.

16. Frosh, "Face of Television," 95.

17. Sarah A. Birnie and Peter Horvath, "Psychological Predictors of Internet Social Communication," *Journal of Computer-Mediated Communication* 7, no. 4 (2002): https://academic.oup.com/jcmc/article/7/4/JCMC743/4584232; Janet Morahan-Martin and Phyllis Schumacher, "Loneliness and Social Uses of the Internet," *Computers in Human Behavior* 19, no. 6 (2003): 659–71; Gill Valentine, "Globalizing Intimacy: The Role of Information and Communication Technologies in Maintaining and Creating Relationships," *Women's Studies Quarterly* 34, nos. 1–2 (2006): 365–93.

18. Lauren Berlant and Michael Warner, "Public Sex," *Critical Inquiry* 24, no. 2 (Winter 1998): 547–66.

19. YouTube, 9 October 2018, https://www.youtube.com/

20. Kate Eichhorn, "Archival Genres: Gathering Texts and Reading Spaces," *Invisible Culture: An Electronic Journal for Visual Culture* 12 May 2008, 12 September 2021, http://www.rochester.edu/in_visible_culture/Issue_12/eichhorn/index.htm#_edn2

21. YouTube, "Policies," 22 July 2019, https://www.youtube.com/yt/about/policies/#community-guidelines

22. YouTube, "Community Guidelines," 3 October 2016, https://www.youtube.com/yt/policyandsafety/communityguidelines.html

23. YouTube, "About YouTube," 13 January 2017, https://www.youtube.com/yt/about/

24. Jean Burgess and Joshua Green, *YouTube: Online Video and Participatory Culture* (Cambridge, UK: Polity Press, 2013).

25. Andrew Tolson, "A New Authenticity? Communicative Practices on YouTube," *Critical Discourse Studies* 7, no. 4 (2010): 286.

26. Robert Vianello, "The Power Politics of 'Live' Television," *Journal of Film and Video* 37, no. 3 (Summer 1985): 26–40.

27. Emma Maguire, "Self-Branding, Hotness, and Girlhood in the Video Blogs of Jenna Marbles," *Biography* 38, no. 1 (2015): 72–86.

28. Harry Cheadle, "What is ASMR? That Good Tingly Feeling No One Can Explain," *VICE*, 31 July 2012, 9 October 2018, https://www.vice.com/en_us/article /gqww3j/asmr-the-good-feeling-no-one-can-explain; ASMR University, 3 February 2021, https://asmruniversity.com/tag/asmr-research-org/

29. Jenn Allen, as cited in Harry Cheadle, "What is ASMR? That Good Tingly Feeling No One Can Explain," *VICE*, 31 July 2012, 9 October 2018, https://www.vice .com/en_us/article/gqww3j/asmr-the-good-feeling-no-one-can-explain

30. Giulia Poerio, "Could Insomnia Be Relieved with a YouTube Video? The Relaxation and Calm of ASMR," in *The Restless Compendium: Interdisciplinary Investigations of Rest and Its Opposites*, ed. Felicity Callard, Kimberley Staines, and James Wilkes (Cham, Switzerland: Palgrave Macmillan, 2016), 119–28.

31. ASMR University, "What is ASMR?" 26 January 2019, https://asmruniversity .com/about-asmr/what-is-asmr/

32. TouchingTingles, "Long ASMR video ~Tapping on an I Pod~ (Long nails)," YouTube, 31 March 2013, 4 September 2018, https://www.youtube.com/watch ?v=hXwzwYIQl5A

33. Glory ASMR, "♦ Cell Phone Tapping Sounds with Long Nails ♦ Soft Spoken About Discovering ASMR ♦ w/ Sticky Plastic," YouTube, 14 August 2016, 11 August 2018, https://www.youtube.com/watch?v=XXin_PWKq2E

34. LJMJBI, "ASMR 1H OF TAPPING NO TALKING! more than 1 hour!" 2 September 2018, https://www.youtube.com/watch?v=WVyPFXdn0ME

35. Michael Andor Brodeur, "Inside the 'Whispering Video' World of ASMR," *Boston Globe*, 29 November 2014, 22 October 2018, https://www.bostonglobe.com/arts /2014/11/29/inside-world-asmr/zxWiAiMXVG8mkis2x1mHUJ/story.html

36. GentleWhispering, as cited in Brodeur, "Inside the 'Whispering Video' World of ASMR."

37. Jordan Pearson, "Inside the Roleplay Subculture Delivering Tingling 'Braingasms' on YouTube," *VICE*, 30 July 2014, 1 July 2021, https://www.vice.com/en /article/3dkapn/inside-the-roleplay-subculture-delivering-tingling-braingasms-on -youtube

38. ASMR Shortbread, "Making You Feel SO Good ♥ [ASMR] ~ hand movements, face touching," YouTube, 24 June 2021, 19 July 2021, https://www.youtube.com /watch?v=ppDHqgeH7SU

39. Joceline Andersen, "Now You've Got the Shiveries: Affect, Intimacy, and the ASMR Whisper Community," *Television and New Media* 16, no. 8 (2015): 691.

40. Steven Connor, *The Book of Skin* (London: Reaktion Books, 2004); Tarja Laine, "Cinema as Second Skin," *New Review of Film and Television Studies* 4, no. 2 (2006): 93–106.

41. Diana Adis Tahhan, "Touching at Depth: The Potential of Feeling and Connection," *Emotion, Space and Society* 7 (2013): 45–53.

42. Sara Ahmed, "The Skin of the Community: Affect and Boundary Formation," in *Revolt, Affect, Collectivity: The Unstable Boundaries of Kristeva's Polis*, ed. Tina Chanter and Ewa Plonowska Ziarek (Albany: State University of New York Press, 2012), 95–111; Nicolette Bragg, "'Beside Myself': Touch, Maternity and the Question of Embodiment," *Feminist Theory* 21, no. 2 (2020): 141–55.

43. Gentle Whispering ASMR, as cited in Julie Beck, "How to Have a 'Brain Orgasm,'" *Atlantic*, 16 December 2013, 31 August 2018, https://www.theatlantic.com/health /archive/2013/12/how-to-have-a-brain-orgasm/282356/

44. Laura Mulvey, *Visual and Other Pleasures* (Bloomington: Indiana University Press, 1989).

45. Tasha Bjelić, "Digital Care," *Women and Performance: A Journal of Feminist Theory* 26, no. 1 (2016): 102.

46. Andersen, "Now You've Got," 685.

47. Tahhan, "Touching at Depth."

48. ASMR Bakery, "ASMR iPhone & iPad Tapping (No Talking)," YouTube, 15 May 2018, 21 August 2018, https://www.youtube.com/watch?v=59IwdmfyI6U

49. bee, "ASMR iPhone & iPad Tapping (No Talking)," YouTube, 21 August 2018, https://www.youtube.com/watch?v=59IwdmfyI6U

50. Yvon Bonenfant, "On Sound and Pleasure: Meditations on the Human Voice," *Sounding Out!* 30 June 2014, 29 August 2018, https://soundstudies.wordpress.com /2014/06/30/on-sound-and-pleasure-meditations-on-the-human-voice/?iframe=true &preview=true

51. Tap Asmr, "Play games with me (nail tapping) | ASMR," YouTube, 7 September 2017, 6 September 2018, https://www.youtube.com/watch?v=KIMufHMS9VU&t=14s

52. Nanna Verhoeff, "Theoretical Consoles: Concepts for Gadget Analysis," *Journal of Visual Culture* 8, no. 3 (2009): 279–98.

53. Sees Nails, "ASMR Tapping On My Phone With My Long NATURAL Nails," YouTube, 19 August 2015, 13 August 2018, https://www.youtube.com/watch?v=cYiFXaM0JVk

54. Sees Nails, "About," YouTube, 17 October 2018, https://www.youtube.com/user/seesnails/about

55. Ella, "ASMR Tapping On My Phone With My Long NATURAL Nails," YouTube, 13 August 2018, https://www.youtube.com/watch?v=cYiFXaM0JVk

56. Bill M, "ASMR Tapping On My Phone With My Long NATURAL Nails," YouTube, 13 August 2018, https://www.youtube.com/watch?v=cYiFXaM0JVk

57. For discussions of the ways emoji amplify affect, see Monica A. Riordan, "Emojis as Tools for Emotion Work: Communicating Affect in Text Messages," *Journal of Language and Social Psychology* 36, no. 5 (2017): 549–67; Luke Stark and Kate Crawford, "The Conservatism of Emoji: Work, Affect, and Communication," *Social Media + Society* (July–December 2015): 1–11.

58. thatASMRchick, "[ASMR] First ASMR video: Tapping/scratching cell phone, hand movements, brushing the camera." YouTube, 26 November 2012, 13 August 2018, https://www.youtube.com/watch?v=yTTkm5ogtfk

59. ASMR Shortbread, "Making You Feel SO Good ♥ [ASMR] ~ hand movements, face touching," YouTube, 24 June 2021, 19 July 2021, https://www.youtube.com/watch?v=ppDHqgeH7SU

60. Joshua Hudelson, "Listening to Whisperers: Performance, ASMR Community and Fetish on YouTube," *Sounding Out!* 10 December 2012, 29 August 2018, https://soundstudiesblog.com/2012/12/10/whisper-community/

61. Gentle Whispering ASMR, "•••Tapping just in Case••• ASMR Soft Spoken Gentle Tapping," YouTube, 16 September 2015, 11 October 2018, https://www.youtube.com/watch?v=Wb-kQMYwU8w

62. Solfrid ASMR, "ASMR Whisper What's On My iPhone | Nail Tapping," YouTube, 2 March 2018, 11 October 2018, https://www.youtube.com/watch?v=nX-hppurSBM

63. Louis Althusser, "Ideology and Ideological State Apparatuses," in *The Anthropology of the State: A Reader*, ed. Aradhana Sharma and Akhil Gupta (Malden, MA: Blackwell Publishing, 2006), 86–111.

64. Peter J. Capuano, *Changing Hands: Industry, Evolution, and the Reconfiguration of the Victorian Body* (Ann Arbor: University of Michigan Press, 2015); Tom Tyler, "The Rule of Thumb," *JAC* 30, nos. 3–4 (2010): 435–56.

65. Sees Nails, "ASMR Tapping On My Phone With My Long NATURAL Nails," YouTube, 19 August 2015, 13 August 2018, https://www.youtube.com/watch?v=cYiFXaM0JVk

66. Jocie B ASMR, "ASMR natural nails ✿ simple care routine | tapping, flutters, whispers," YouTube, 8 August 2019, 1 December 2020, https://www.youtube.com

/watch?v=D6qYQdyronY; Nanou ASMR, "ASMR TAPPING WITH FAKE NAILS!" YouTube, 19 April 2020, 1 December 2020, https://www.youtube.com/watch?v=Mi BZLBbl4nM

67. TheRedBaron Lives! "[ASMR] Fast Screen Tapping," YouTube, 21 August 2018, https://www.youtube.com/watch?v=jL2PNwLjL4I

68. Ellawyn ASMR, "Tapping and Scratching with Long and Short Nails ASMR," YouTube, 16 August 2018, 9 September 2018, https://www.youtube.com/watch ?v=0fJNFSU1rAg

69. ASMR StitchesScritches, "ASMR Sunday Shortie: lightly binaural tapping, long and short nails, fast and slow, no talking," YouTube, 28 September 2014, 9 September 2018, https://www.youtube.com/watch?v=cep7oWlIW78

70. Stitches Scritches ASMR, "FAQ - Stitches Scritches ASMR," Tumblr, 15 October 2018, http://stitchescritchesasmr.tumblr.com/FAQ

71. ASMR StitchesScritches, "ASMR Sunday Shortie: lightly binaural tapping, long and short nails, fast and slow, no talking," YouTube, 28 September 2014, 9 September 2018, https://www.youtube.com/watch?v=cep7oWlIW78

72. quietexperiment, "ASMR Tapping & Scratching 6: Cell Phones of History!" YouTube, 21 March 2012, 12 August 2018, https://www.youtube.com/watch ?v=rhGA5aSJAgg

73. Cheeks ASMR, "ASMR Old Cell Phone Tapping/Button Pressing - No Talking," YouTube, 29 July 2016, 4 September 2018, https://www.youtube.com/watch?v=A9y Le8Pq5OA

74. ahmadaamer6, "ASMR Old Cell Phone Tapping/Button Pressing - No Talking," YouTube, 4 September 2018, https://www.youtube.com/watch?v=A9yLe8Pq5OA

75. Ann Cvetkovich, "Photographing Objects as Queer Archival Practice," in *Feeling Photography*, ed. Elspeth H. Brown and Thy Phu (Durham, NC: Duke University Press, 2014), 273–96.

76. Ann Cvetkovich, as cited in Tammy Rae Carland and Ann Cvetkovich, "Sharing an Archive of Feelings: A Conversation," *Art Journal* 72, no. 2 (Summer 2013): 73.

77. Ann Cvetkovich, *An Archive of Feelings: Trauma, Sexuality, and Lesbian Public Cultures* (Durham, NC: Duke University Press, 2003).

78. Of course, "older media" and "old media" are also new media. When these technologies were introduced, they were often identified as new, and people were less likely to be familiar with how to use them because there were no histories of established engagements. See Benjamin Peters, "And Lead Us Not into Thinking the New Is New: A Bibliographic Case for New Media History," *New Media and Society* 11, nos. 1–2 (2009): 13–30.

79. Quiet Time ASMR, "ASMR PHONE TAPPING," YouTube, 5 December 2017, 6 September 2018, https://www.youtube.com/watch?v=O3STnD6oG_8

80. MaricoL13, "ASMR PHONE TAPPING," YouTube, 6 September 2018, https://www.youtube.com/watch?v=O3STnD6oG_8

81. Alladin Zilva, "Texting with long nails," YouTube, 11 August 2018, https://www.youtube.com/watch?v=KWZMg470uWI

82. Stephen Arcella, "Binaural ASMR Long Nails Tapping on Cellphone Case," YouTube, 11 August 2018, https://www.youtube.com/watch?v=juiNlRmAXKc

83. alsrg, "ASMR Tapping on Ipad with Long Nails ♥ | Tingly Gentle Tapping (No talking!)," YouTube, 1 September 2018, https://www.youtube.com/watch?v=gSpEPt1SsGE; Chiara ASMR, "ASMR Tapping on Ipad with Long Nails ♥ | Tingly Gentle Tapping (No talking!)," YouTube, 1 September 2018, https://www.youtube.com/watch?v=gSpEPt1SsGE

84. ibokki ASMR, "LEAVE YOUR ASMR REQUESTS," YouTube, 22 November 2017, 21 October 2018, https://www.youtube.com/watch?v=6dqxwmci2tM

85. Clareee ASMR, "ASMR Doing YOUR Requests!! - Assortment," YouTube, 20 September 2018, 21 October 2018, https://www.youtube.com/watch?v=LYYzz0FymNw

86. Gibi ASMR, "[ASMR] Top Requested Triggers ~," YouTube, 19 April 2018, 21 October 2018, https://www.youtube.com/watch?v=iUo1JXnlRXY

87. Cvetkovich, as cited in Carland and Cvetkovich, "Sharing an Archive," 76.

88. Peace and Saraity ASMR, "ASMR| Tapping on Cellphone Cases| iPhone X Case Collection," YouTube, 2 January 2018, 22 October 2018, https://www.youtube.com/watch?v=hwUknNkXiyU

89. Jordan Maxwell, "ASMR| Tapping on Cellphone Cases| iPhone X Case Collection," YouTube, 3 February 2022, https://www.youtube.com/watch?v=hwUknNkXiyU

90. Emma E, "ASMR iPhone & iPad Tapping (No Talking)," YouTube, 21 August 2018, https://www.youtube.com/watch?v=59IwdmfyI6U

91. Kayla Livingston, "ASMR| Tapping on Cellphone Cases| iPhone X Case Collection," YouTube, 22 October 2018, https://www.youtube.com/watch?v=hwUknNkXiyU

92. Mike Sanchez, "Binaural ASMR Long Nails Tapping on Cellphone Case," YouTube, 11 August 2018, https://www.youtube.com/watch?v=juiNlRmAXKc

93. Daryk, "ASMR 1H OF TAPPING NO TALKING! more than 1 hour!" YouTube, 2 September 2018, https://www.youtube.com/watch?v=WVyPFXdn0ME

94. Aero, "ASMR 🎧 TAPPING (& SCRATCHING) 📱 mobile phone + how I write on touch screen with long natural nails 👀," YouTube, 21 August 2018, https://www.youtube.com/watch?v=KDXh7hxmKUk

95. careful aesthetic, "ASMR 1H OF TAPPING NO TALKING! more than 1 hour!" YouTube, 2 September 2018, https://www.youtube.com/watch?v=WVyPFXdn0ME

96. Charlotte, "ASMR Tapping on Phone No Talking," YouTube, 4 September 2018, https://www.youtube.com/watch?v=529LnRAF_gk

97. Jairo L., "ASMR ~ tapping & scratching ~ (Long nails)," YouTube, 2 September 2018, https://www.youtube.com/watch?v=fT7EpmFGpO8

98. Ayla ASMR, "ASMR≈♥ Delicate Tapping ♥ iPad and iPhone || AylaASMR," YouTube, 29 January 2016, 1 September 2018, https://www.youtube.com/watch?v=JVnt7sqFIPo

99. Barthes, *Camera Lucida*.

100. Emojipedia, "Heart Hands," 18 March 2022, https://emojipedia.org/heart-hands/

101. Solfrid ASMR, "ASMR Whisper What's On My iPhone | Nail Tapping," YouTube, 2 March 2018, 11 October 2018, https://www.youtube.com/watch?v=nX-hppurSBM

102. goodpigASMR, "ASMR≈♥ Delicate Tapping ♥ iPad and iPhone || AylaASMR," YouTube, 1 September 2018, https://www.youtube.com/watch?v=JVnt7sqFIPo

103. Ayla ASMR, "ASMR≈♥ Delicate Tapping ♥ iPad and iPhone || AylaASMR," YouTube, 1 September 2018, https://www.youtube.com/watch?v=JVnt7sqFIPo

104. Eden Cameron, "ASMR Pure Varied Tapping ~ Wood, Glass, Phone, Leather, 3Dio Metal Parts & Ears (NO TALKING) 2 Hours," YouTube, 23 August 2018, https://www.youtube.com/watch?v=jbvxlvvrLuU

105. skinny legend, "ASMR 1H OF TAPPING NO TALKING! more than 1 hour!" YouTube, 2 September 2018, https://www.youtube.com/watch?v=WVyPFXdn0ME

106. BTS Trash, "ASMR 1H OF TAPPING NO TALKING! more than 1 hour!" YouTube, 2 September 2018, https://www.youtube.com/watch?v=WVyPFXdn0ME

107. Althusser, "Ideology and Ideological State Apparatuses."

Afterword

1. Squeaker Tweeker, "Social distancing anxiety," Urban Dictionary, 21 March 2020, 19 September 2020, https://www.urbandictionary.com/define.php?term=Social+distancing+anxiety

2. Raj Persaud and Peter Bruggen, "How People Panicked Over the Pandemic," *Psychology Today*, 16 September 2020, https://www.psychologytoday.com/us/blog/slightly-blighty/202009/how-people-panicked-over-the-pandemic; Reginald D. Williams, Arnav Shah, Roosa Tikkanen, Eric C. Schneider, and Michelle M. Doty, "Do

Americans Face Greater Mental Health and Economic Consequences from COVID-19? Comparing the U.S. with Other High-Income Countries," The Commonwealth Fund, 6 August 2020, 19 September 2020, https://www.commonwealthfund.org /publications/issue-briefs/2020/aug/americans-mental-health-and-economic -consequences-COVID19; John W. Ayers, Eric C. Leas, Derek C. Johnson, Adam Poliak, Benjamin M. Althouse, Mark Dredze, and Alicia L. Nobles, "Internet Searches for Acute Anxiety During the Early Stages of the COVID-19 Pandemic," *JAMA Internal Medicine*, 24 August 2020, 19 September 2020, https://jamanetwork.com /journals/jamainternalmedicine/fullarticle/2769543

3. John M. Grohol, "Coronavirus Anxiety: Social Distancing Helps Stop the Spread," PsychCentral, 9 April 2020, 19 September 2020, https://psychcentral.com/blog /coronavirus-anxiety-social-distancing-helps-stop-the-spread/

4. Centers for Disease Control and Prevention (CDC), "Social Distancing, Quarantine, and Isolation," 15 July 2020, 19 September 2020, https://www.cdc.gov /coronavirus/2019-ncov/prevent-getting-sick/social-distancing.html

5. Ali Pattillo, "The most common social distancing confusion, explained by medical experts," Inverse, 17 September 2020, https://www.inverse.com/mind-body /social-distancing-coronavirus-survey

6. ABC7 Chicago, "Man knocked unconscious during social distancing fight at Burbank buffet restaurant," 17 September 2020, https://abc7chicago.com/social -distancing-fight-two-volcanoes-burbank/6432527/; Alex Costello, "Man Fractured Another's Skull Over Social Distancing: Nassau PD," NY Patch, 19 September 2020, https://patch.com/new-york/freeport/man-fractured-anothers-skull-over-social -distancing-nassau-pd

7. Vittoria Traverso, "How social-distancing symbols are changing our cities," BBC, 16 September 2020, https://www.bbc.com/worklife/article/20200909-how-social -distancing-symbols-are-changing-our-cities

8. Gartholomew83, "social distancing," Urban Dictionary, 12 March 2020, 19 September 2020, https://www.urbandictionary.com/define.php?term=Social%20distancing

9. Kaydog1, "social distancing," Urban Dictionary, 23 March 2020, 19 September 2020, https://www.urbandictionary.com/define.php?term=Social%20distancing. The italic emphasis appears in Kaydog1's text.

10. NoThankYou4PiggyFlu, "social distancing," Urban Dictionary, 1 May 2009, 19 September 2020, https://www.urbandictionary.com/define.php?term=Social%20distancing

11. The italic emphasis appears in NoThankYou4PiggyFlu's text.

12. Chris Stokel-Walker, "How personal contact will change post-Covid-19," BBC, 29 April 2020, 29 June 2020, https://www.bbc.com/future/article/20200429-will -personal-contact-change-due-to-coronavirus

13. Eve Kosofsky Sedgwick, *Touching Feeling: Affect, Pedagogy, Performativity* (Durham, NC: Duke University Press, 2003), 17.

14. Evan L. Ardiel and Catharine H. Rankin, "The Importance of Touch in Development," *Paediatrics and Child Health* 15, no. 3 (2010): 153–56; Elizabeth Bush, "The Use of Human Touch to Improve the Well-being of Older Adults: A Holistic Nursing Intervention," *Journal of Holistic Nursing* 19, no. 3 (2001): 256–70.

15. Eric T. Lehman, "'Washing Hands, Reaching Out'—Popular Music, Digital Leisure and Touch During the COVID-19 Pandemic," *Leisure Sciences* 23, nos. 1–2 (2021): 273–79.

16. Lehman, "'Washing Hands, Reaching Out,'" 275.

17. Martin S. Pernick, "Contagion and Culture," *American Literary History* 14, no. 4 (2002): 858.

18. Diana Adis Tahhan, "Touching at Depth: The Potential of Feeling and Connection," *Emotion, Space and Society* 7 (2013), 45–53.

19. Amanda Holpuch, "'We can't go back to normal': The People Left Behind in America's Covid Recovery," *Guardian*, 14 August 2021, 10 September 2021, https://www.theguardian.com/inequality/2021/aug/14/women-fighting-covid-recovery-fair

20. Meditative Lullaby ASMR, "ASMR 🫁 Deep Breathing Affirmations Meditation + Healing Music for Coronavirus / COVID-19 Anxiety," YouTube, 13 April 2020, 28 May 2020, https://www.youtube.com/watch?v=LcYWjP-LZiw

21. Ozley ASMR, "ASMR Calming You Down (corona virus edition)," YouTube, 17 March 2020, 21 May 2020, https://www.youtube.com/watch?v=tQ0hNVFgryg

22. Madison Phoenix ASMR, "ASMR MASSAGE RP 💆 Ear Massage Lotion & Cupping [Coronavirus Anxiety Relief]," YouTube, 18 March 2020, 25 May 2020, https://www.youtube.com/watch?v=ABIyOswHQCc

23. Gibi ASMR, "ASMR | Touching & Tapping The Camera Lens," YouTube, 10 October 2019, 27 May 2020, https://www.youtube.com/watch?v=x-gOV7GmM5U

24. Bunny Marthy, "Coronavirus ASMR - #ASMR #Relaxing 46/100," YouTube, 19 March 2020, 25 May 2020, https://www.youtube.com/watch?v=nuGhTxQJecw

25. Perry The platypus, "ASMR | Touching & Tapping The Camera Lens," YouTube, 27 May 2020, https://www.youtube.com/watch?v=x-gOV7GmM5U

26. Halunke Nr. Eins, "ASMR | Touching & Tapping The Camera Lens," YouTube, 27 May 2020, https://www.youtube.com/watch?v=x-gOV7GmM5U

27. Seafoam Kitten's ASMR, "ASMR - YOU ARE MY CAMERA ~ Personal Attention and Affection! Lens Touching, Button Tapping ~," YouTube, 1 February 2019, 27 May 2020, https://www.youtube.com/watch?v=2oWT40AVpCM

28. Gibi ASMR, "ASMR | Touching & Tapping The Camera Lens," YouTube, 10 October 2019, 27 May 2020, https://www.youtube.com/watch?v=x-gOV 7GmM5U

29. Geoffrey A. Fowler and Heather Kelly, "'Screen Time' Has Gone from Sin to Survival Tool," *Washington Post*, 9 April 2020, 29 June 2020, https://www .washingtonpost.com/technology/2020/04/09/screen-time-rethink-coronavirus/

30. speed wagon, "ASMR | Touching & Tapping The Camera Lens," YouTube, 27 May 2020, https://www.youtube.com/watch?v=x-gOV7GmM5U

31. Louise Anne A., "ASMR | Touching & Tapping The Camera Lens," YouTube, 27 May 2020, https://www.youtube.com/watch?v=x-gOV7GmM5U

32. Spitfire11511, "ASMR | Touching & Tapping The Camera Lens," YouTube, 22 September 2020, https://www.youtube.com/watch?v=x-gOV7GmM5U

33. Aldo gamas, "ASMR | Touching & Tapping The Camera Lens," YouTube, 22 September 2020, https://www.youtube.com/watch?v=x-gOV7GmM5U

34. Stanley Mikko Santiago, "ASMR Calming You Down (corona virus edition)," YouTube, 21 May 2020, https://www.youtube.com/watch?v=tQ0hNVFgryg

35. Sc O, "ASMR Calming You Down (corona virus edition)," YouTube, 21 May 2020, https://www.youtube.com/watch?v=tQ0hNVFgryg

36. Tim lovely, "CORONAVIRUS ASMR ~ Breathing- and Affirmation Exercises To Manage Anxiety," YouTube, 25 May 2020, https://www.youtube.com/watch ?v=vCAjxfpjmEA

37. Gerard Zandvliet, "CORONAVIRUS ASMR ~ Breathing- and Affirmation Exercises To Manage Anxiety," YouTube, 25 May 2020, https://www.youtube.com/watch ?v=vCAjxfpjmEA

38. ASMR SWEETIE, "CORONAVIRUS ASMR ~ Breathing- and Affirmation Exercises To Manage Anxiety," YouTube, 25 May 2020, https://www.youtube.com/watch ?v=vCAjxfpjmEA

39. HeatheredEffect ASMR, "ASMR | Real Talk about Covid 19 -- (Comforting Whispered Ramble)," YouTube, 15 April 2020, 28 May 2020, https://www.youtube.com /watch?v=6PUEHjXzipY

40. Squeaker Tweeker, "Social distancing anxiety," Urban Dictionary, 21 March 2020, 19 September 2020, https://www.urbandictionary.com/define.php?term=Social +distancing+anxiety

41. Seafoam Kitten's ASMR, "ASMR - YOU ARE MY CAMERA ~ Personal Attention and Affection! Lens Touching, Button Tapping ~," YouTube, 1 February 2019, 25 September 2020, https://www.youtube.com/watch?v=2oWT40AVpCM

42. Bruh, "ASMR - YOU ARE MY CAMERA ~ Personal Attention and Affection! Lens Touching, Button Tapping ~," YouTube, 27 May 2020, https://www.youtube.com /watch?v=2oWT40AVpCM

43. nmspy, "ASMR - YOU ARE MY CAMERA ~ Personal Attention and Affection! Lens Touching, Button Tapping ~," YouTube, 27 May 2020, https://www.youtube .com/watch?v=2oWT40AVpCM

44. Samurai LAN, "ASMR - YOU ARE MY CAMERA ~ Personal Attention and Affection! Lens Touching, Button Tapping ~," YouTube, 27 May 2020, https://www .youtube.com/watch?v=2oWT40AVpCM

45. Roland Barthes, *Camera Lucida: Reflections on Photography*, trans. Richard Howard (New York: Farrar, Strauss, and Giroux, 1981), 10.

46. Centers for Disease Control and Prevention (CDC), "What You Need To Know About Handwashing," YouTube, 16 December 2019, 3 October 2020, https://www .youtube.com/watch?v=d914EnpU4Fo&feature=emb_rel_end

47. Centers for Disease Control and Prevention (CDC), "Clean Hands Save Lives," 3 October 2020, https://www.cdc.gov/handwashing/

48. Derek Thompson, "Hygiene Theater Is a Huge Waste of Time," *Atlantic*, 27 July 2020, 28 November 2020, https://www.theatlantic.com/ideas/archive/2020/07 /scourge-hygiene-theater/614599/

49. Ian Carleton Schaefer and Alison E. Gabay, "Rule 3: Wash Your Hands—Return to Work in the Time of COVID-19 [VIDEO]," *National Law Review*, 3 September 2020, 17 September 2020, https://www.natlawreview.com/article/rule-3-wash-your -hands-return-to-work-time-covid-19-video

50. Stephanie Newell, *Histories of Dirt: Media and Urban Life in Colonial and Postcolonial Lagos* (Durham, NC: Duke University Press, 2019).

51. Sheryl N. Hamilton, "Hands in Cont(r)act: The Resiliency of Business Handshakes in Pandemic Culture," *Canadian Journal of Law & Society/La Revue Canadienne Droit et Société* 34, no. 2 (2019): 354.

52. Jenny Gross, "How to Stop Touching Your Face," *New York Times*, 5 March 2020, 21 June 2020, https://www.nytimes.com/2020/03/05/health/stop-touching-your-face -coronavirus.html

53. Natasha Piñon, "Worried About Coronavirus? Stop Touching Your Face," Mashable, 17 September 2020, https://in.mashable.com/science/11827/worried-about -coronavirus-stop-touching-your-face

54. Ingrid Banks, *Hair Matters: Beauty, Power, and Black Women's Consciousness* (New York and London: New York University Press, 2000).

55. Kalhan Rosenblatt, "Try not to touch your face. Also, try not to think about touching your face," NBC News, 5 March 2020, 22 September 2020, https://www .nbcnews.com/tech/social-media/try-not-touch-your-face-also-try-not-think-about -n1150161

56. Jeffrey D. Fisher, Marvin Rytting, and Richard Heslin, "Hands Touching Hands: Affective and Evaluative Effects of an Interpersonal Touch," *Sociometry* (1976): 416–21; Alberto Gallace and Charles Spence, "The Science of Interpersonal Touch: An Overview," *Neuroscience and Biobehavioral Reviews* 34, no. 2 (2010): 246–59; Judith Anne Horton, Pauline Rose Clance, Claire Sterk-Elifson, and James Emshoff, "Touch in Psychotherapy: A Survey of Patients' Experiences," *Psychotherapy: Theory, Research, Practice, Training* 32, no. 3 (1995): 443–57.

57. Cláudia Simão and Beate Seibt, "Friendly Touch Increases Gratitude by Inducing Communal Feelings," *Frontiers in Psychology* 6 (June 2015): 1–11.

58. Jeroen Camps, Chloé Tuteleers, Jeroen Stouten, and Jill Nelissen, "A Situational Touch: How Touch Affects People's Decision Behavior," *Social Influence* 8, no. 4 (2013): 237–50.

59. Kathryn Graham, Sharon Bernards, Antonia Abbey, Tara M. Dumas, and Samantha Wells, "When Women Do Not Want It: Young Female Bargoers' Experiences with and Responses to Sexual Harassment in Social Drinking Contexts," *Violence Against Women* 23, no. 12 (2017): 1419–41.

60. Raksha Kumar, "#MeToo in the age of coronavirus," LiveMint, 1 May 2020, 22 July 2020, https://www.livemint.com/mint-lounge/features/-metoo-in-the-age-of -coronavirus-11588304252814.html

61. Elisa Martinuzzi, "When Work Moved Home during Covid, so Did Toxic Workplace Harassment," *Daily Herald*, 21 June 2020, https://www.dailyherald.com /business/20200621/analysis-as-work-has-moved-home-so-has-harassment; Louise Esola, "Telecommuting puts anti-harassment measures in focus," Business Insurance, 27 March 2020, 22 July 2020, https://www.businessinsurance.com/article /20200327/NEWS06/912333751/Telecommuting-puts-anti-harassment-measures-in -focus-coronavirus-COVID-19-pandem#; Sian Norris and Claudia Torrisi, "COVID-19 hasn't killed sexual harassment at work – it's just moved online," openDemocracy, 23 September 2020, https://www.opendemocracy.net/en/5050/covid19-sexual -harassment-work-online/

62. Melissa Jeltsen, "Home Is Not a Safe Place for Everyone," *HuffPost*, 13 March 2020, 22 July 2020, https://www.huffpost.com/entry/domestic-violence-coronavirus _n_5e6a6ac1c5b6bd8156f3641b?ncid=engmodushpmg00000004

63. Jilly Boyce Kay, "'Stay the fuck at home!' Feminism, Family and the Private Home in a Time of Coronavirus," *Feminist Media Studies* 20, no. 6 (2020): 883.

64. Jonathan Chadwick, "Handshakes and Public Hugs Could Go Extinct in Human Society When the Coronavirus Pandemic Ends, Scientists Warn," *Daily Mail*, 30 March 2020, 16 September 2020, https://www.dailymail.co.uk/sciencetech/article -8167051/Handshakes-public-hugs-extinct-pandemic-ends-scientists-warn.html

65. Robert Dingwall, as cited in Chadwick, "Handshakes and Public Hugs Could Go Extinct."

66. Star-Ledger Editorial Board, "'I'm Still Terrified.' Will #MeToo end Trenton's Culture of Silence? A Q&A," 19 August 2020, 16 September 2020, https://www.nj .com/opinion/2020/08/im-still-terrified-will-metoo-end-trentons-culture-of-silence -a-qa.html

67. See Michele White, *Producing Masculinity: The Internet, Gender, and Sexuality* (New York: Routledge, 2019).

68. Nickie D. Phillips and Nicholas Chagnon, "Where's the Panic, Where's the Fire? Why Claims of Moral Panic and Witch Hunts Miss the Mark When It Comes to Campus Rape and MeToo," *Feminist Media Studies* 21, no. 3 (2021): 409.

69. Kristin J. Anderson, Melinda Kanner, and Nisreen Elsayegh, "Are Feminists Man Haters? Feminists' and Nonfeminists' Attitudes Toward Men," *Psychology of Women Quarterly* 33 (2009): 216–24.

70. Cathrine Jansson-Boyd, "Coronavirus is accelerating a culture of no touching – here's why that's a problem," The Conversation, 16 March 2020, 22 August 2021, https://theconversation.com/coronavirus-is-accelerating-a-culture-of-no-touching -heres-why-thats-a-problem-133488

71. Carol Kinsey Goman, "Coronavirus calls for social distancing and an end to shaking hands. Here's what we're losing." NBC News, 12 March 2020, 21 September 2021, https://www.nbcnews.com/think/opinion/coronavirus-calls-social-distancing -end-shaking-hands-here-s-what-ncna1157136

72. Mandy Oaklander, "The Coronavirus Killed the Handshake and the Hug. What Will Replace Them?" *Time*, 27 May 2020, 21 June 2020, https://time.com/5842469 /coronavirus-handshake-social-touch/

73. Rosemary Clark-Parsons, "'I SEE YOU, I BELIEVE YOU, I STAND WITH YOU': #MeToo and the Performance of Networked Feminist Visibility," *Feminist Media Studies* 21, no. 3 (2021): 362.

74. Carrie A. Rentschler, "Bystander Intervention, Feminist Hashtag Activism, and the Anti-carceral Politics of Care," *Feminist Media Studies* 17, no. 4 (2017): 565–84.

75. Carey-Ann Morrison, "Heterosexuality and Home: Intimacies of Space and Spaces of Touch," *Emotion, Space and Society* 5 (2012): 15.

76. Ann J. Cahill, "Foucault, Rape, and the Construction of the Feminine Body," *Hypatia* 15, no. 1 (2000): 56.

77. Kari Stefansen, "Understanding Unwanted Sexual Touching: A Situational Approach," in *Rape in the Nordic Countries: Continuity and Change*, ed. Marie Bruvik Heinskou, May-Len Skilbrei, and Kari Stefansen (London: Routledge, 2020), 63.

78. Jeff Link, "Will We Ever Want to Use Touchscreens Again?: The future is touchless." Built In, 24 June 2020, https://builtin.com/design-ux/future-touchless

79. Christian Hetrick, "No, Coronavirus Won't Kill the Touchscreen," *Philadelphia Inquirer*, 15 June 2020, 26 June 2020, https://www.inquirer.com/consumer/coronavirus-touchscreen-kiosks-20200615.html

80. Lisa Bonos, "'I would pay $50 for a 2-minute hug': True Stories of Sex Starvation during Self-quarantine," *Washington Post*, 16 April 2020, 21 June 2020, https://www.washingtonpost.com/lifestyle/2020/04/16/sex-touch-quarantine-lockdown-social-distance-coronavirus/?arc404=true

81. Erika Hughes, as cited in Francesca Gillett, "Coronavirus: Why going without physical touch is so hard," BBC News, 25 April 2020, 20 May 2020, https://www.bbc.com/news/uk-52279411

82. Marina Koren, "The New Cringeworthy," *Atlantic*, 17 April 2020, 22 July 2020, https://www.theatlantic.com/science/archive/2020/04/coronavirus-pandemic-cringe/610180/

83. Muriel Dimen, "Sexuality and Suffering, or the Eew! Factor," *Studies in Gender and Sexuality* 6, no.1 (2005): 1–18.

84. Julia Kristeva, *The Powers of Horror: An Essay on Abjection* (New York: Columbia University Press, 1982).

85. Shanley Pierce, "Touch starvation is a consequence of COVID-19's physical distancing," TMC News, 15 May 2020, https://www.tmc.edu/news/2020/05/touch-starvation/

86. Laura Mulvey, *Visual and Other Pleasures* (Bloomington: Indiana University Press, 1989).

87. Sara Ahmed, *Living a Feminist Life* (Durham, NC: Duke University Press, 2017).

Selected Bibliography

Ahmed, Sara. "Affective Economies." *Social Text* 22, no. 2 (79) (Summer 2004): 117–39.

Ahmed, Sara. *Living a Feminist Life*. Durham, NC: Duke University Press, 2017.

Ahmed, Sara. "The Skin of the Community: Affect and Boundary Formation." In *Revolt, Affect, Collectivity: The Unstable Boundaries of Kristeva's Polis*, edited by Tina Chanter and Ewa Plonowska Ziarek, 95–111. Albany: State University of New York Press, 2012.

Akrich, Madeleine. "The De-scription of Technical Objects." In *Shaping Technology/ Building Society: Studies in Sociotechnical Change*, edited by Wiebe E. Bijker and John Law, 205–24. Cambridge, MA: MIT Press, 1994.

Albert, Noel, and Dwight Merunka. "The Role of Brand Love in Consumer-Brand Relationships." *Journal of Consumer Marketing* 30, no. 3 (2013): 258–66.

Althusser, Louis. "Ideology and Ideological State Apparatuses." In *The Anthropology of the State: A Reader*, edited by Aradhana Sharma and Akhil Gupta, 86–111. Malden, MA: Blackwell Publishing, 2006.

Altieri, Charles. "Affect, Intentionality, and Cognition: A Response to Ruth Leys." *Critical Inquiry* 38 (Summer 2012): 878–81.

Andersen, Joceline. "Now You've Got the Shiveries: Affect, Intimacy, and the ASMR Whisper Community." *Television and New Media* 16, no. 8 (2015): 683–700.

Anderson, Kristin J., Melinda Kanner, and Nisreen Elsayegh. "Are Feminists Man Haters? Feminists' and Nonfeminists' Attitudes Toward Men." *Psychology of Women Quarterly* 33 (2009): 216–24.

Anzieu, Didier. *The Skin-ego*, translated by Naomi Segal. London: Karnac Books, 2016.

Ardiel, Evan L., and Catharine H. Rankin. "The Importance of Touch in Development." *Paediatrics and Child Health* 15, no. 3 (2010): 153–56.

Arxer, Steven L. "Hybrid Masculine Power: Reconceptualizing the Relationship Between Homosociality and Hegemonic Masculinity." *Humanity and Society* 35 (November 2011): 390–422.

Bakhtin, Mikhail. *Rabelais and His World*, translated by Hélène Iswolsky. Bloomington: Indiana University Press, 1984.

Banet-Weiser, Sarah. *Empowered: Popular Feminism and Popular Misogyny*. Durham, NC: Duke University Press, 2018.

Banet-Weiser, Sarah, and Kate M. Miltner. "#MasculinitySoFragile: Culture, Structure, and Networked Misogyny." *Feminist Media Studies* 16, no. 1 (2016): 171–74.

Banks, Ingrid. *Hair Matters: Beauty, Power, and Black Women's Consciousness*. New York and London: New York University Press, 2000.

Barthes, Roland. *Camera Lucida: Reflections on Photography*, translated by Richard Howard. New York: Farrar, Strauss, and Giroux, 1981.

Barthes, Roland. *Incidents*, translated by Richard Howard. Berkeley: University of California Press, 1992.

Bartky, Sandra L. "Foucault, Femininity, and the Modernization of Patriarchal Power." In *Feminism and Foucault*, edited by Irene Diamond and Lee Quinby, 61–86. Boston: Northeastern University Press, 1988.

Berlant, Lauren, and Michael Warner. "Public Sex." *Critical Inquiry* 24, no. 2 (Winter 1998): 547–66.

Birnie, Sarah A., and Peter Horvath. "Psychological Predictors of Internet Social Communication." *Journal of Computer-Mediated Communication* 7, no. 4 (2002): https://academic.oup.com/jcmc/article/7/4/JCMC743/4584232

Bitman, Nomy, and Nicholas A. John. "Deaf and Hard of Hearing Smartphone Users: Intersectionality and the Penetration of Ableist Communication Norms." *Journal of Computer-Mediated Communication* 24 (2019): 56–72.

Bittman, Michael, James Mahmud Rice, and Judy Wajcman. "Appliances and Their Impact: The Ownership of Domestic Technology and Time Spent on Household Work." *British Journal of Sociology* 55, no. 3 (2004): 401–23.

Bjelić, Tasha. "Digital Care." *Women and Performance: A Journal of Feminist Theory* 26, no. 1 (2016): 101–4.

Bonenfant, Yvon. "On Sound and Pleasure: Meditations on the Human Voice." *Sounding Out!* (30 June 2014): https://soundstudies.wordpress.com/2014/06/30/on-sound-and-pleasure-meditations-on-the-human-voice/?iframe=true&preview=true

Bragg, Nicolette. "'Beside Myself': Touch, Maternity and the Question of Embodiment." *Feminist Theory* 21, no. 2 (2020): 141–55.

Bratich, Jack. "The Digital Touch: Craft-work as Immaterial Labour and Ontological Accumulation." *Ephemera: Theory and Politics in Organization* 10, nos. 3–4 (2010): 303–18.

Brothers Grimm. *The Complete Grimm's Fairy Tales*, translated by Margaret Hunt. New York: Pantheon Books, 1944.

Brown, Elspeth H., and Thy Phu. "Introduction." In *Feeling Photography*, edited by Elspeth H. Brown and Thy Phu, 1–28. Durham, NC: Duke University Press, 2014.

Brownmiller, Susan. *Femininity*. New York: Simon and Schuster, 1984.

Budgeon, Shelley. "Individualized Femininity and Feminist Politics of Choice." *European Journal of Women's Studies*, 22, no. 3 (2015): 303–18.

Burgess, Jean, and Joshua Green. *YouTube: Online Video and Participatory Culture*. Cambridge, UK: Polity Press, 2013.

Bush, Elizabeth. "The Use of Human Touch to Improve the Well-being of Older Adults: A Holistic Nursing Intervention." *Journal of Holistic Nursing* 19, no. 3 (2001): 256–70.

Cahill, Ann J. "Foucault, Rape, and the Construction of the Feminine Body." *Hypatia* 15, no. 1 (2000): 43–63.

Camps, Jeroen, Chloé Tuteleers, Jeroen Stouten, and Jill Nelissen. "A Situational Touch: How Touch Affects People's Decision Behavior." *Social Influence* 8, no. 4 (2013): 237–50.

Capuano, Peter J. *Changing Hands: Industry, Evolution, and the Reconfiguration of the Victorian Body*. Ann Arbor: University of Michigan Press, 2015.

Carland, Tammy Rae, and Ann Cvetkovich. "Sharing an Archive of Feelings: A Conversation." *Art Journal* 72, no. 2 (Summer 2013): 70–77.

Carroll, Barbara A., and Aaron C. Ahuvia. "Some Antecedents and Outcomes of Brand Love." *Marketing Letters* 17 (2006): 79–89.

Chatzidakis, Andreas, Jamie Hakim, Jo Littler, Catherine Rottenberg, and Lynne Segal. "From Carewashing to Radical Care: The Discursive Explosions of Care During Covid-19." *Feminist Media Studies* 20, no. 6 (2020): 889–95.

Clark-Parsons, Rosemary. "'I SEE YOU, I BELIEVE YOU, I STAND WITH YOU': #MeToo and the Performance of Networked Feminist Visibility." *Feminist Media Studies* 21, no. 3 (2021): 362–80.

Clover, Carol. *Men, Women, and Chain Saws: Gender in the Modern Horror Film*. Princeton, NJ: Princeton University Press, 1992.

Connor, Steven. *The Book of Skin*. London: Reaktion Books, 2004.

Cooley, Heidi Rae. "It's All About the Fit: The Hand, the Mobile Screenic Device and Tactile Vision." *Journal of Visual Culture* 3, no. 2 (2004): 133–55.

Cowan, Ruth Schwartz. *More Work for Mother: The Ironies of Household Technology from the Open Hearth to the Microwave*. New York: Basic Books, 1983.

Cranny-Francis, Anne. "Semefulness: A Social Semiotics of Touch." *Social Semiotics* 21, no. 4 (September 2011): 463–81.

Cromby, John, and Martin E. H. Willis. "Affect—or Feeling (After Leys)." *Theory and Psychology* 26, no. 4 (2016): 476–95.

Culler, Jonathan. *Literary Theory: A Very Short Introduction*. New York: Oxford University Press, 1997.

Cvetkovich, Ann. *An Archive of Feelings: Trauma, Sexuality, and Lesbian Public Cultures*. Durham, NC: Duke University Press, 2003.

Cvetkovich, Ann. "Photographing Objects as Queer Archival Practice." In *Feeling Photography*, edited by Elspeth H. Brown and Thy Phu, 273–96. Durham, NC: Duke University Press, 2014.

Derrida, Jacques. *On Touching—Jean-Luc Nancy*, translated by Christine Irizarry. Stanford, CA: Stanford University Press, 2005.

Dimen, Muriel. "Sexuality and Suffering, or the Eew! Factor." *Studies in Gender and Sexuality* 6, no.1 (2005): 1–18.

Doane, Mary Ann. *Femmes Fatales: Feminism, Film Theory, Psychoanalysis*. New York: Routledge, 1991.

Douglas, Mary. *Implicit Meanings: Selected Essays in Anthropology*. London: Routledge, 1999.

Dux, Monica, and Zora Simic. *The Great Feminist Denial*. Melbourne: Melbourne University Press, 2008.

Dyer, Richard. "The Colour of Virtue: Lillian Gish, Whiteness, and Femininity." In *Women and Film: A Sight and Sound Reader*, edited by Pam Cook and Philip Dodd, 1–9. Philadelphia: Temple University Press, 1993.

Eagleton, Terry. *How to Read a Poem*. Malden, MA: Blackwell Publishing, 2007.

Ehrenreich, Barbara, and Deirdre English. *For Her Own Good: Two Centuries of the Experts' Advice to Women*, 2nd ed. New York: Random House, 2005.

Eichhorn, Kate. "Archival Genres: Gathering Texts and Reading Spaces." *Invisible Culture: An Electronic Journal for Visual Culture* 12 (May 2008): http://www.rochester .edu/in_visible_culture/Issue_12/eichhorn/index.htm#_edn2

Ellis, John. *Visible Fictions: Cinema: Television: Video*. London: Routledge, 2002.

Farrell, Amy Erdman. *Fat Shame: Stigma and the Fat Body in American Culture*. New York: New York University Press, 2011.

Feuer, Jane. *The Hollywood Musical*, 2nd edition. Bloomington: Indiana University Press, 1993.

Fisher, Jeffrey D., Marvin Rytting, and Richard Heslin. "Hands Touching Hands: Affective and Evaluative Effects of an Interpersonal Touch." *Sociometry* (1976): 416–21.

Frank, Elena. "Groomers and Consumers: The Meaning of Male Body Depilation to a Modern Masculinity Body Project." *Men and Masculinities* 17, no. 3 (2014): 278–98.

Freud, Sigmund. "Femininity." In *New Introductory Lectures on Psycho-analysis*, translated by James Strachey, 139–67. New York: W. W. Norton, 1965.

Frosh, Paul. "The Face of Television." *Annals of the American Academy* 625 (September 2009): 87–102.

Fuchs, Christoph, Martin Schreier, and Stijn M. J. van Osselaer. "The Handmade Effect: What's Love Got to Do with It?" *Journal of Marketing* 79 (March 2015): 98–110.

Gallace, Alberto, and Charles Spence. "The Science of Interpersonal Touch: An Overview." *Neuroscience and Biobehavioral Reviews* 34, no. 2 (2010): 246–59.

Ging, Debbie, Theodore Lynn, and Pierangelo Rosati. "Neologising Misogyny: Urban Dictionary's Folksonomies of Sexual Abuse." *New Media and Society* 22, no. 5 (2020): 838–56.

Ginsberg, Elaine, and Sara Lennox. "Antifeminism in Scholarship and Publishing." In *Antifeminism in the Academy*, edited by Veve Clark, Shirley Nelson Garner, Margaret Higonnet, and Ketu Katrak, 169–99. New York: Routledge, 2014.

Gitelman, Lisa, and Geoffrey B. Pingree. "What's New About New Media?" In *New Media, 1740–1915*, edited by Lisa Gitelman and Geoffrey B. Pingree, xi–xxii. Cambridge, MA: MIT Press, 2003.

Glover, David, and Cora Kaplan. *Genders*. London: Routledge, 2000.

Goggin, Gerard. "Disability and Haptic Mobile Media." *New Media and Society* 19, no. 10 (2017): 1563–80.

Goggin, Gerard, and Christopher Newell. *Digital Disability*. Lanham, MD: Rowman and Littlefield, 2003.

Graham, Kathryn, Sharon Bernards, Antonia Abbey, Tara M. Dumas, and Samantha Wells. "When Women Do Not Want It: Young Female Bargoers' Experiences with and Responses to Sexual Harassment in Social Drinking Contexts." *Violence Against Women* 23, no. 12 (2017): 1419–41.

Grosz, Elizabeth. *Volatile Bodies: Toward a Corporeal Feminism*. Bloomington: Indiana University Press, 1994.

Hamilton, Sheryl N. "Hands in Cont(r)act: The Resiliency of Business Handshakes in Pandemic Culture." *Canadian Journal of Law and Society/La Revue Canadienne Droit et Société* 34, no. 2 (2019): 343–60.

Haraway, Donna. *Simians, Cyborgs, and Women: The Reinvention of Nature*. New York: Routledge, 1991.

Hayles, N. Katherine. *How We Think: Digital Media and Contemporary Technogenesis*. Chicago: University of Chicago Press, 2012.

Hayles, N. Katherine. "Hyper and Deep Attention: The Generational Divide in Cognitive Modes." *Profession* (2007): 187–99.

Heidegger, Martin. *What Is Called Thinking?* translated by J. Glenn Gray. New York: HarperCollins Publishers, 1976.

Hérubel, Jean-Pierre V. M. "The Darker Side of Light: Heidegger and Nazism: A Bibliographic Essay." *Shofar* 10, no. 1 (Fall 1991): 85–105.

Hilmes, Michele. "The Television Apparatus: Direct Address." *Journal of Film and Video* 37, no. 4 (Fall 1985): 27–36.

Hirshman, Linda R. *Get to Work: A Manifesto for Women of the World*. New York: Viking, 2006.

Horton, Judith Anne, Pauline Rose Clance, Claire Sterk-Elifson, and James Emshoff. "Touch in Psychotherapy: A Survey of Patients' Experiences." *Psychotherapy: Theory, Research, Practice, Training* 32, no. 3 (1995): 443–57.

Hudelson, Joshua. "Listening to Whisperers: Performance, ASMR Community and Fetish on YouTube." *Sounding Out!* (10 December 2012): https://soundstudiesblog.com/2012/12/10/whisper-community/

Johnson, Barbara. "Teaching Deconstructively." In *Writing and Reading Differently*, edited by George Douglas Atkins, Michael L. Johnson, and Nancy R. Comley, 140–48. Lawrence: University of Kansas Press, 1986.

Kay, Jilly Boyce. "'Stay the fuck at home!' Feminism, Family and the Private Home in a Time of Coronavirus." *Feminist Media Studies* 20, no. 6 (2020): 883–88.

Keating, Patrick. *Hollywood Lighting from the Silent Era to Film Noir*. New York: Columbia University Press, 2009.

Keller, Jessalynn. "Beware the Dancing Communist: Alexandria Ocasio-Cortez, Snaplash, and the Politics of Embodied Joy." In *Anti-feminisms in Media Culture*, edited by Michele White and Diane Negra, 158–75. London: Routledge, 2022.

Kirschenbaum, Matthew. *Mechanisms: New Media and the Forensic Imagination*. Cambridge, MA: MIT Press, 2008.

Kristeva, Julia. *The Powers of Horror: An Essay on Abjection*. New York: Columbia University Press, 1982.

Lacan, Jacques. *The Four Fundamental Concepts of Psycho-analysis*, translated by A. Sheridan. New York: W. W. Norton, 1981.

Lægran, Anne Sofie, and James Stewart. "Nerdy, Trendy, or Healthy? Configuring the Internet Café." *New Media and Society* 5, no. 3 (2003): 357–77.

Lafrance, Marc. "Skin Studies: Past, Present and Future." *Body and Society* 24, nos. 1–2 (2018): 3–32.

Laine, Tarja. "Cinema as Second Skin." *New Review of Film and Television Studies* 4, no. 2 (2006): 93–106.

Langer, John. "Television's 'Personality System.'" *Media, Culture and Society* 3, no. 4 (1981): 351–65.

Leet, Elizabeth S. "Objectification, Empowerment, and the Male Gaze in the Lanval Corpus." *Historical Reflections* 42, no. 1 (Spring 2016): 75–87.

Lehman, Eric T. "'Washing Hands, Reaching Out'—Popular Music, Digital Leisure and Touch During the COVID-19 Pandemic." *Leisure Sciences* 43, nos. 1–2 (2021): 273–79.

Levine, Philippa, and Alison Bashford. "Introduction: Eugenics and the Modern World." In *The Oxford Handbook of the History of Eugenics*, edited by Alison Bashford and Philippa Levine, 3–24. New York: Oxford University Press, 2010.

Leys, Ruth. "The Turn to Affect: A Critique." *Critical Inquiry* 37 (Spring 2011): 434–72.

Lombard, Matthew, and Jennifer Snyder-Duch. "Interactive Advertising and Presence." *Journal of Interactive Advertising* 1, no. 2 (2001): 56–65.

Losh, Elizabeth. *Hashtag*. New York: Bloomsbury Publishing, 2019.

Maguire, Emma. "Self-Branding, Hotness, and Girlhood in the Video Blogs of Jenna Marbles." *Biography* 38, no. 1 (2015): 72–86.

Marks, Laura U. *The Skin of the Film: Intercultural Cinema, Embodiment, and the Senses*. Durham, NC: Duke University Press, 2000.

Marks, Laura U. *Touch: Sensuous Theory and Multisensory Media*. Minneapolis: University of Minnesota Press, 2002.

Marks, Laura U. "Video Haptics and Erotics." *Screen* 39, no. 4 (1998): 331–47.

Marvin, Carolyn. *When Old Technologies Were New: Thinking About Electric Communication in the Late Nineteenth Century*. New York: Oxford University Press, 1988.

Marzullo, Michelle, Jasmine Rault, and T. L. Cowan. "Can I Study You? Cross-Disciplinary Studies in Queer Internet Studies." *First Monday* 23, no. 7 (2 July 2018): http://www.firstmonday.dk/ojs/index.php/fm/article/view/9263/7465

Massanari, Adrienne. "#Gamergate and The Fappening: How Reddit's Algorithm, Governance, and Culture Support Toxic Technocultures." *New Media and Society* 19, no. 3 (2017): 329–46.

Matamoros-Fernández, Ariadna. "Inciting Anger Through Facebook Reactions in Belgium: The Use of Emoji and Related Vernacular Expressions in Racist Discourse." *First Monday* 23, no. 9 (3 September 2018): https://www.firstmonday.org/ojs/index.php/fm/article/view/9405

McCormack, M. "The Declining Significance of Homohysteria for Male Students in Three Sixth Forms in the South of England." *British Educational Research Journal* 37, no. 2 (April 2011): 37–53.

McHugh, Kevin E. "Touch at a Distance: Toward a Phenomenology of Film." *Geo-Journal* 80 (2015): 839–51.

Merleau-Ponty, Maurice. *Signs*, translated by Richard C. McCleary. Evanston, IL: Northwestern University Press, 1964.

Metz, Christian. *The Imaginary Signifier: Psychoanalysis and the Cinema*, translated by Celia Britton, Annwyl Williams, Ben Brewster, and Alfred Guzzetti. Bloomington: Indiana University Press, 1982.

Miller, J. Hillis. "Derrida Enisled." *Critical Inquiry* 33, no. 2 (Winter 2007): 248–76.

Moores, Shaun. "Digital Orientations: 'Ways of the Hand' and Practical Knowing in Media Uses." *Mobile Media and Communication* 2, no. 2 (2014): 196–208.

Morahan-Martin, Janet, and Phyllis Schumacher. "Loneliness and Social Uses of the Internet." *Computers in Human Behavior* 19, no. 6 (2003): 659–71.

Morrison, Carey-Ann. "Heterosexuality and Home: Intimacies of Space and Spaces of Touch." *Emotion, Space and Society* 5 (2012): 10–18.

Mulvey, Laura. *Visual and Other Pleasures*. Bloomington: Indiana University Press, 1989.

Muñiz, Albert M., and Thomas C. O'Guinn. "Brand Community." *Journal of Consumer Research* 27, no. 4 (March 2001): 412–32.

Murray, Sarah. "Postdigital Cultural Studies." *International Journal of Cultural Studies* 23, no. 4 (2020): 441–50.

Nead, Lynda. "Seductive Canvases: Visual Mythologies of the Artist and Artistic Creativity." *Oxford Art Journal* 18, no. 2 (1995): 59–69.

Newell, Stephanie. *Histories of Dirt: Media and Urban Life in Colonial and Postcolonial Lagos*. Durham, NC: Duke University Press, 2019.

Nochlin, Linda. *Women, Art, and Power and Other Essays*. New York: Harper and Row, 1984.

Oudshoorn, Nelly, and Trevor Pinch. *How Users Matter: The Co-construction of Users and Technologies*. Cambridge, MA: MIT Press, 2003.

Oudshoorn, Nelly, Els Rommes, Marcelle Stienstra. "Configuring the User as Everybody: Gender and Design Cultures in Information and Communication Technologies." *Science, Technology, and Human Values* 29, no. 1 (Winter 2004): 30–63.

Paasonen, Susanna. "A Midsummer's Bonfire: Affective Intensities of Online Debate." In *Networked Affect*, edited by Ken Hillis, Susanna Paasonen, and Michael Petit, 27–42. Cambridge, MA: MIT Press, 2015.

Paasonen, Susanna, Ken Hillis, and Michael Petit. "Introduction: Networks of Transmission: Intensity, Sensation, Value." In *Networked Affect*, edited by Ken Hillis, Susanna Paasonen, and Michael Petit, 1–25. Cambridge, MA: MIT Press, 2015.

Parisi, David. *Archaeologies of Touch: Interfacing with Haptics from Electricity to Computing*. Minneapolis: University of Minnesota Press, 2018.

Patrick, Adele. "Defiantly Dusty: A (Re)Figuring of 'Feminine Excess.'" *Feminist Media Studies* 1, no. 3 (2001): 361–78.

Pernick, Martin S. "Contagion and Culture." *American Literary History* 14, no. 4 (2002): 858–65.

Phillips, Nickie D., and Nicholas Chagnon. "Where's the Panic, Where's the Fire? Why Claims of Moral Panic and Witch Hunts Miss the Mark When It Comes to Campus Rape and MeToo." *Feminist Media Studies* 21, no. 3 (2021): 409–26.

Pile, Steve. "Spatialities of Skin: The Chafing of Skin, Ego and Second Skins in T. E. Lawrence's Seven Pillars of Wisdom." *Body and Society* 17, no. 4: (2011): 57–81.

Pink, Sarah, Jolynna Sinanan, Larissa Hjorth, and Heather Horst. "Tactile Digital Ethnography: Researching Mobile Media Through the Hand." *Mobile Media and Communication* 4, no. 2 (2016): 237–51.

Plotnick, Rachel. "Force, Flatness and Touch Without Feeling: Thinking Historically About Haptics and Buttons." *New Media and Society* 19, no. 10 (2017): 1632–52.

Poerio, Giulia. "Could Insomnia Be Relieved with a YouTube Video? The Relaxation and Calm of ASMR." In *The Restless Compendium: Interdisciplinary Investigations of Rest and Its Opposites*, edited by Felicity Callard, Kimberley Staines, James Wilkes, 119–28. Cham, Switzerland: Palgrave Macmillan, 2016.

Pollock, Griselda. *Vision and Difference: Femininity, Feminism and the Histories of Art.* London: Routledge, 1988.

Qiu, Lin, Han Lin, Angela K. Leung, and William Tov. "Putting Their Best Foot Forward: Emotional Disclosure on Facebook." *Cyberpsychology, Behavior, and Social Networking* 15, no. 10 (2012): 569–72.

Rasmussen, Majken Kirkegaard, and Marianne Graves Petersen. "Re-scripting Interactive Artefacts with Feminine Values." In *Proceedings of the 2011 Conference on Designing Pleasurable Products and Interfaces*, 1–8. Milan: ACM, 2011.

Rentschler, Carrie A. "Bystander Intervention, Feminist Hashtag Activism, and the Anti-carceral Politics of Care." *Feminist Media Studies* 17, no. 4 (2017): 565–84.

Richard, Craig. *Brain Tingles: The Secret to Triggering Autonomous Sensory Meridian Response for Improved Sleep, Stress Relief, and Head-to-Toe Euphoria.* New York: Simon and Schuster, 2018.

Richardson, Shaun, and Amy Donley. "'That's So Ghetto!' A Study of the Racial and Socioeconomic Implications of the Adjective Ghetto." *Theory in Action* 11, no. 4 (October 2018): 22–43.

Riordan, Monica A. "Emojis as Tools for Emotion Work: Communicating Affect in Text Messages." *Journal of Language and Social Psychology* 36, no. 5 (2017): 549–67.

Roberts, Kevin. *Lovemarks: The Future Beyond Brands.* New York: PowerHouse Books, 2004.

Rommes, Els. "Creating Places for Women on the Internet: The Design of a 'Women's Square' in a Digital City." *European Journal of Women's Studies* 9, no. 4 (2002): 400–29.

Rommes, Els, Ellen van Oost, and Nelly Oudshoorn. "Gender in the Design of the Digital City of Amsterdam." *Information, Communication, and Society* 2, no. 4 (1999): 476–95.

Ruihley, Brody J., and Joshua R. Pate. "For the Love of Sport: Examining Sport Emotion Through a Lovemarks Lens." *Communication and Sport* 5, no. 2 (2017): 135–59.

Sammond, Nicholas. *Birth of an Industry: Blackface Minstrelsy and the Rise of American Animation.* Durham, NC: Duke University Press, 2015.

Sas, Corina, Alan Dix, Jennefer Hart, and Ronghui Su. "Dramaturgical Capitalization of Positive Emotions: The Answer for Facebook Success?" In *Proceedings of the 23rd British HCI Group Annual Conference on People and Computers: Celebrating People and Technology*, 120–29. London: British Computer Society, 2009.

Schumann, Jennifer, Sandrine Zufferey, and Steve Oswald. "What Makes a Straw Man Acceptable? Three Experiments Assessing Linguistic Factors." *Journal of Pragmatics* 141 (2019): 1–15.

Sedgwick, Eve Kosofsky. *Touching Feeling: Affect, Pedagogy, Performativity*. Durham, NC: Duke University Press, 2003.

Segal, Naomi. *Consensuality: Didier Anzieu, Gender and the Sense of Touch*. Amsterdam: Rodopi, 2009.

Segal, Naomi. "'A Petty Form of Suffering': A Brief Cultural Study of Itching." *Body and Society* 24, nos. 1–2 (2018): 88–102.

Seigworth, Gregory J., and Melissa Gregg. "An Inventory of Shimmers." In *The Affect Theory Reader*, edited by Melissa Gregg and Gregory J. Seigworth, 1–26. Durham, NC: Duke University Press, 2010.

Shade, Leslie Regan. "Feminizing the Mobile: Gender Scripting of Mobiles in North America." *Continuum: Journal of Media and Cultural Studies* 21, no. 2 (June 2007): 179–89.

Sherry, John F., and Mary Ann McGrath. "Unpacking the Holiday Presence: A Comparative Ethnography of Two Gift Stores." *Interpretive Consumer Research* (1989): 148–67.

Shukin, Nicole. "The Hidden Labour of Reading Pleasure." *English Studies in Canada* 33, nos. 1–2 (March–June 2007): 23–27.

Simão, Cláudia, and Beate Seibt. "Friendly Touch Increases Gratitude by Inducing Communal Feelings." *Frontiers in Psychology* 6 (June 2015): 1–11.

Smith, Herbert. "Badges, Buttons, T-shirts and Bumperstickers: The Semiotics of Some Recursive Systems." *Journal of Popular Culture* 21, no. 4 (1988): 141–49.

Smith, Shawn Michelle. "Photography Between Desire and Grief: Roland Barthes and F. Holland Day." In *Feeling Photography*, edited by Elspeth H. Brown and Thy Phu, 29–46. Durham, NC: Duke University Press, 2014.

Sobchack, Vivian. "What My Fingers Knew: The Cinesthetic Subject, or Vision in the Flesh." *Senses of Cinema* 5 (2000): http://sensesofcinema.com/2000/conference -special-effects-special-affects/fingers/

Sobieraj, Sarah, and Jeffrey M. Berry. "From Incivility to Outrage: Political Discourse in Blogs, Talk Radio, and Cable News." *Political Communication* 28, no. 1 (2011): 19–41.

Stark, Luke, and Kate Crawford. "The Conservatism of Emoji: Work, Affect, and Communication." *Social Media + Society* (July–December 2015): 1–11.

Stefansen, Kari. "Understanding Unwanted Sexual Touching: A Situational Approach." In *Rape in the Nordic Countries: Continuity and Change*, edited by Marie Bruvik Heinskou, May-Len Skilbrei, and Kari Stefansen, 49–65. London: Routledge, 2020.

Stein, Jordan Alexander. *Avidly Reads Theory*. New York: New York University Press, 2019.

Sung, Kenzo K. "'Hella Ghetto!' (Dis)locating Race and Class Consciousness in Youth Discourses of Ghetto Spaces, Subjects and Schools." *Race Ethnicity and Education*, 18, no. 3 (2015): 363–95.

Tahhan, Diana Adis. "Touching at Depth: The Potential of Feeling and Connection." *Emotion, Space and Society 7* (2013): 45–53.

Talisse, Robert, and Scott F. Aikin. "Two Forms of the Straw Man." *Argumentation* (2006): 345–52.

Thrift, Nigel. "Closer to the Machine? Intelligent Environments, New Forms of Possession and the Rise of the Supertoy." *Cultural Geographies* 10 (2003): 389–407.

Tolson, Andrew. "A New Authenticity? Communicative Practices on YouTube." *Critical Discourse Studies* 7, no. 4 (2010): 277–89.

Tomkins, Silvan S. "Script Theory: Differential Magnification of Affects." In *Nebraska Symposium on Motivation*, edited by H. E. Howe and R. A. Dienstbier, 201–36. Lincoln: University of Nebraska Press, 1978.

Tyler, Tom. "The Rule of Thumb." *JAC* 30, nos. 3–4 (2010): 435–56.

Valentine, Gill. "Globalizing Intimacy: The Role of Information and Communication Technologies in Maintaining and Creating Relationships." *Women's Studies Quarterly* 34, nos. 1–2 (2006): 365–93.

Van den Brink, Marieke, and Yvonne Benschop. "Gender Practices in the Construction of Academic Excellence: Sheep with Five Legs." *Organization* 19, no. 4 (2012): 507–24.

Verhoeff, Nanna. "Theoretical Consoles: Concepts for Gadget Analysis." *Journal of Visual Culture* 8, no. 3 (2009): 279–98.

Vianello, Robert. "The Power Politics of 'Live' Television." *Journal of Film and Video* 37, no. 3 (Summer 1985): 26–40.

Vinken, P. J. *The Shape of the Heart.* Amsterdam: Elsevier, 2000.

Viser, Victor J. "Mode of Address, Emotion and Stylistics: Images of Children in American Magazine Advertising, 1940–1950." *Communication Research* 24, no. 1 (February 1997): 83–101.

Waldron, Emma Leigh. "'This FEELS SO REAL!' Sense and Sexuality in ASMR Videos." *First Monday* 22, nos. 1–2 (January 2017): https://firstmonday.org/ojs/index .php/fm/article/view/7282/5804

West, Jevin D., Jennifer Jacquet, Molly M. King, Shelley J. Correll, and Carl T. Bergstrom. "The Role of Gender in Scholarly Authorship." *PLoS One* 8, no. 7 (2013): 1–6.

White, Michele. *The Body and the Screen: Theories of Internet Spectatorship.* Cambridge, MA: MIT Press, 2006.

White, Michele. *Buy It Now: Lessons from eBay*. Durham, NC: Duke University Press, 2012.

White, Michele. "GIFs from Feminists: Visual Pleasure, Danger, and Anger on the Jezebel Website." *Feminist Formations* 30, no. 2 (Summer 2018): 202–30.

White, Michele. "How 'your hands look' and 'what they can do': #ManicureMonday, Twitter, and Useful Media." *Feminist Media Histories* 1, no. 2 (Spring 2015): 4–36.

White, Michele. *Producing Masculinity: The Internet, Gender, and Sexuality*. New York: Routledge, 2019.

White, Michele. *Producing Women: The Internet, Traditional Femininity, Queerness, and Creativity*. New York: Routledge, 2015.

White, Michele. "Representations or People?" *Ethics and Information Technology* 4, no. 3 (2002): 249–66.

White, Michele. "Television and Internet Differences by Design: Rendering Liveness, Presence, and Lived Space." *Convergence: The International Journal of Research into New Media Technologies* 12, no. 3 (August 2006): 341–55.

White, Michele. "Women's Nail Polish Blogging and Femininity: 'The Girliest You Will Ever See Me.'" In *Cupcakes, Pinterest, and Ladyporn: Feminized Popular Culture in the Early Twenty-First Century*, edited by Elana Levine, 137–56. Urbana-Champaign: University of Illinois Press, 2015.

Woolgar, Steve. "Configuring the User: The Case of Usability Trials." In *A Sociology of Monsters: Essays on Power, Technology and Domination*, edited by John Law, 58–99. London: Routledge, 1991.

Index